UNITED
NATIONS
PEACEKEEPING
and the
NON-USE OF FORCE

APPLIED BAYESIAN FORECASTING and TIME SERIES ANALYSIS

TEXTS IN STATISTICAL SCIENCE SERIES

Editors:

Dr. Chris Chatfield
Reader in Statistics, School of Mathematical Sciences, University of Bath, UK

Professor Jim V. Zidek,
Department of Statistics, University of British Columbia, Canada

OTHER TITLES IN THE SERIES INCLUDE:

Practical Statistics for Medical Research
D.G. Altman

Interpreting Data
A.J.B. Anderson

Statistical Methods for SPC and TQM
D. Bissell

Statistics in Research and Development
Second edition
R. Caulcutt

The Analysis of Time Series
Fourth edition
C. Chatfield

Problem Solving—A Statisticians Guide
C. Chatfield

Introduction to Multivariate Analysis
C. Chatfield and A. J. Collins

Modelling Binary Data
D. Collett

Modelling Survival Data in Medical Research
D. Collett

Applied Statistics
D. R. Cox and E. J. Snell

Statistical Analysis of Reliability Data
M. J. Crowder, A. C. Kimber, T. J. Sweeting and R. L. Smith

An Introduction to Generalized Linear Models
A. J. Dobson

Introduction to Optimization Methods and their Applications in Statistics
B. S. Everitt

Multivariate Statistics—A Practical Approach
B. Flury and H. Riedwyl

Readings in Decision Analysis
S. French

Multivariate Analysis of Variance and Repeated Measures
D. J. Hand and C. C. Taylor

The Theory of Linear Models
B. Jørgensen

Statistics for Accountants
S. Letchford

Statistical Theory
Fourth Edition
B. Lindgren

Randomization and Monte Carlo Methods in Biology
B. F. J. Manly

Statistical Methods in Agriculture and Experimental Biology
Second edition
R. Mead, R. N. Curnow and A. M. Hasted

Statistics in Engineering
A. V. Metcalfe

Elements of Simulation
B. J. T. Morgan

Probability: Methods and Measurement
A. O'Hagan

Essential Statistics
Second edition
D. G. Rees

Large Sample Methods in Statistics
P. K. Sen and J. M. Singer

Decision Analysis: A Bayesian Approach
J. Q. Smith

Applied Nonparametric Statistical Methods
Second edition
P. Sprent

Elementary Applications of Probability Theory
Second edition
H. C. Tuckwell

Statistical Process Control: Theory and Practice
Third edition
G. B. Wetherill and D. W. Brown

APPLIED BAYESIAN FORECASTING and TIME SERIES ANALYSIS

Andy Pole
Statistical Analyst
Blackhawk Limited Partnership

Mike West
Professor of Statistics
Institute of Statistics and Decision Sciences, Duke University

Jeff Harrison
Professor of Statistics
University of Warwick, UK

CHAPMAN & HALL/CRC
Boca Raton London New York Washington, D.C.

Published in 1994 by
Chapman & Hall/CRC
Taylor & Francis Group
6000 Broken Sound Parkway NW, Suite 300
Boca Raton, FL 33487-2742

© 1994 by Taylor & Francis Group, LLC
Chapman & Hall/CRC is an imprint of Taylor & Francis Group

No claim to original U.S. Government works
Printed in the United States of America on acid-free paper
10 9 8 7 6 5 4

International Standard Book Number-10: 0-412-04401-3 (Hardcover)
International Standard Book Number-13: 978-0-412-04401-4 (Hardcover)
Library of Congress catalog number: 94-13542

The software mentioned in this book is now available for download on our Web site at:
http://www.crcpress.com/e_products/downloads/default.asp

This book contains information obtained from authentic and highly regarded sources. Reprinted material is quoted with permission, and sources are indicated. A wide variety of references are listed. Reasonable efforts have been made to publish reliable data and information, but the author and the publisher cannot assume responsibility for the validity of all materials or for the consequences of their use.

No part of this book may be reprinted, reproduced, transmitted, or utilized in any form by any electronic, mechanical, or other means, now known or hereafter invented, including photocopying, microfilming, and recording, or in any information storage or retrieval system, without written permission from the publishers.

Trademark Notice: Product or corporate names may be trademarks or registered trademarks, and are used only for identification and explanation without intent to infringe.

Library of Congress Cataloging-in-Publication Data

Catalog record is available from the Library of Congress

Taylor & Francis Group
is the Academic Division of Informa plc.

The software mentioned in this book is now available for download on our Web site at:
http://www.crcpress.com/e_products/downoads/default.asp

Visit the Taylor & Francis Web site at
http://www.taylorandfrancis.com

and the CRC Press Web site at
http://www.crcpress.com

To Our Families

Contents

Preface   xv

Part A: DYNAMIC BAYESIAN MODELLING
Theory and Applications

Chapter 1   Practical Modelling and Forecasting   3
- 1.1   The Nature of Time Series   3
- 1.2   Time Series Analysis vs Forecasting   4
- 1.3   Model Forms   5
- 1.4   Dynamic Systems   6
- 1.5   The Bayesian Approach to Forecasting   9
- 1.6   Future Developments   10
- 1.7   Reference Material   11

Chapter 2   Methodological Framework   13
- 2.1   The Dynamic Linear Model   13
  - 2.1.1   Time and Order   15
- 2.2   Bayesian Analysis   16
  - 2.2.1   Sequential Analysis   16
- 2.3   Subjective Intervention   17
- 2.4   Forecasting   18
- 2.5   Distributional Forms   19
- 2.6   Monitoring   20
- 2.7   Variance Analysis   21
- 2.8   Smoothing   23
- 2.9   Component Models   24
  - 2.9.1   Block Structuring   25
- 2.10   Summary   26
- 2.11   References   27

Chapter 3   Analysis of the DLM   29
- 3.1   Model Form and Notation   29
- 3.2   Updating: Prior to Posterior Analysis   30

        3.2.1   Bayes' Theorem   30
        3.2.2   Prior Information   30
        3.2.3   Forecasts   31
        3.2.4   Likelihood   34
        3.2.5   Posterior Information   34
        3.2.6   Evolution   37
3.3   Forward Intervention   37
        3.3.1   Arbitrary Intervention   38
3.4   Smoothing   40
3.5   Component Forms   42
        3.5.1   Polynomial Trend Components   42
        3.5.2   Seasonal Component Models   44
        3.5.3   Harmonic Analysis   47
        3.5.4   Regression Components   51
3.6   Superposition: Block Structured Models   51
        3.6.1   Block Discounting   52
        3.6.2   Component Intervention   54
3.7   Variance Learning   55
        3.7.1   Variance Laws   58
        3.7.2   Variance Discounting   58
        3.7.3   Variance Intervention   60
        3.7.4   Smoothing   60
3.8   Forecast Monitoring   62
        3.8.1   Bayes Factors   62
        3.8.2   Automatic Monitoring   64
        3.8.3   Cumulative Evidence   64
        3.8.4   Monitor Design   65
        3.8.5   General Monitoring Scheme   65
3.9   Error Analysis   69
3.10   References   69
3.11   Exercises   70

Appendix 3.1   Review of Distribution Theory   75

    Univariate Normal Distribution   75
        Sums of Normal Variables   76
    Multivariate Normal Distribution   76
        Linear Transformations   77
        Marginal Distributions   77
        Conditional Distributions   77
    Gamma Distribution   78
    Univariate Student-t Distribution   78
    Multivariate Student-t Distribution   79

Contents   ix

        Joint Normal-Gamma Distributions   79
                Univariate Normal-Gamma   79
                Multivariate Normal-Gamma   80
        References   80

Appendix 3.2   Classical Time Series Models   83
        Autoregressive Models   83
        Moving Average Models   84
        Autoregressive Moving Average Models   84
        DLM Representation of ARMA Models   84
                Alternative Representations of AR Models   85
        Stationarity   86
        Modelling with ARMA Models   87
        Forecast Function Equivalence of ARMA and DLM   88
        References   89

Chapter 4   Application: Turkey Chick Sales   91
   4.1   Preliminary Investigation   92
        4.1.1   Stabilising Variation   92
        4.1.2   Seasonal Pattern Changes   94
        4.1.3   Forecasting Transformed Series   94
        4.1.4   Assessing Transformations   95
   4.2   Live Forecasting   102
        4.2.1   Forecast Analysis   103
   4.3   Retrospective Perspectives   109
        4.3.1   The View From 1977   110
        4.3.2   The Global Picture   111
        4.3.3   Final Remarks   116
   4.4   Summary   119
   4.5   Exercises   119

Chapter 5   Application: Market Share   121
   5.1   Exploratory Analysis   121
   5.2   Dynamic Regression Model   124
   5.3   A First Analysis   125
        5.3.1   Scaling Regression Variables   127
   5.4   Analysis with Rescaled Promotions   129
        5.4.1   Dynamic Analysis   129
   5.5   Final Analysis   135
   5.6   'What if?' Projections   139
   5.7   Contemporaneous and Lagged Relationships   142

5.8 Multiple Regression Components   142
5.9 Summary   143
5.10 Exercises   143

Chapter 6   Application: Marriages in Greece   147

6.1 Analysis I   148
6.2 Analysis II   151
6.3 Analysis III   154
6.4 Conclusions   156
6.5 1972: Case for the Defence   157
6.6 Regional Analysis   158

   6.6.1 Athens, Macedonia   158
   6.6.2 Epirus, Peloponnesos, Thessaly, Thrace   158
   6.6.3 Rest of Mainland   159
   6.6.4 Crete, Aegean Islands, Ionian Islands   159

6.7 Summary   162
6.8 Exercises   163

Chapter 7   Further Examples and Exercises   165

7.1 Nile River Volume   165
7.2 Gas Consumption   170
7.3 Retail Sales   172
7.4 Inflation   174
7.5 United Kingdom Marriages   175
7.6 Housing Starts   177
7.7 Telephone Calls   180
7.8 Perceptual Speed   182
7.9 Savings   184
7.10 Air Freight   186
7.11 London Mortality   192
7.12 Share Earnings   194
7.13 Passenger Car Sales   198
7.14 Phytoplankton   199
7.15 Further Data Sets   202
7.16 Data Set List   227
7.17 References   230

Contents   xi

Part B: INTERACTIVE TIME SERIES ANALYSIS AND FORECASTING

Chapter 8   Installing BATS   235

8.1   Documentation   235
    8.1.1   User Guide   235
    8.1.2   Reference Guide   236
8.2   How to Use This Guide   236
    8.2.1   Typographical Conventions   237
8.3   Before You Begin   238
8.4   Installing BATS   238
8.5   Using BATS with Floppy Disc Drives   239
8.6   Using BATS with A Fixed Disc Drive   240
8.7   Using Expanded/Extended Memory   241
8.8   Microsoft Windows   241

Chapter 9   Tutorial: Introduction to BATS   243

9.1   Starting a Session   243
9.2   Operating with Menus   243
9.3   Traversing the Menu Hierarchy   245
9.4   Examining Data   246
9.5   Ending a Session   251
9.6   Summary of Operations   252
9.7   Getting out of Trouble   253
9.8   Items Covered in This Tutorial   253

Appendix 9.1   Files and Directories   255

    File Types   255
    File Masks   256
    Directories   257
    The Working Directory   258
    Summary   258

Chapter 10   Tutorial: Introduction to Modelling   259

10.1   Dynamic Modelling   259
10.2   Specifying the Steady Model   260
10.3   Prediction   262
    10.3.1   Changing the Model Dynamic   265
    10.3.2   Modelling Known Variance   266
10.4   Forecasting and Model Estimation   266
    10.4.1   Retrospective Analysis   269
10.5   Summary of Operations   273

- 10.6 Extending the Model: Explanatory Variables 274
  - 10.6.1 Specifying a Regression Component 274
- 10.7 Data Transformations: New Series from Old 280
- 10.8 Prediction with Regressors 282
- 10.9 Multiple Regressions 284
- 10.10 Summary of Operations 285
- 10.11 Extending the Model: Seasonal Patterns 286
  - 10.11.1 Specifying a Seasonal Component 286
  - 10.11.2 Restricted Seasonal Patterns 291
  - 10.11.3 Prediction 295
  - 10.11.4 Prior Specification for Seasonal Components 297
- 10.12 Summary of Operations 298
- 10.13 Items Covered in This Tutorial 299

Chapter 11 Tutorial: Advanced Modelling 301

- 11.1 Data and Model 301
- 11.2 Preliminary Analysis 303
- 11.3 Monitoring Forecast Performance 306
  - 11.3.1 Setting a Monitor 306
  - 11.3.2 Analysis with Monitoring 307
  - 11.3.3 Modes of Automatic Signal Handling 311
  - 11.3.4 Customising the Monitor 312
  - 11.3.5 More Analysis 313
- 11.4 Summary of Operations 316
- 11.5 Intervention Facilities 317
- 11.6 Summary of Operations 324
- 11.7 Forward Intervention 325
  - 11.7.1 Taking the Analysis Further 329
- 11.8 Summary of Operations 330
- 11.9 Putting It All Together 331
- 11.10 Summary of Operations 337
- 11.11 Items Covered in This Tutorial 338
- 11.12 Digression: On the Identification of Change 338
- 11.13 References 340

Chapter 12 Tutorial: Modelling with Incomplete Data 341

- 12.1 Communicating Missing Values 342
- 12.2 Summary of Operations 345
- 12.3 Analysis with Missing Values I: Response Series 346
- 12.4 Analysis with Missing Values II: Regressor Series 349
- 12.5 Summary of Operations 353
- 12.6 Prediction with Missing Regressors 354
- 12.7 Summary of Operations 355

Contents xiii

- 12.8 Data Transformations with Missing Values 355
- 12.9 Summary of Operations 357
- 12.10 Items Covered in This Tutorial 357

Chapter 13   Tutorial: Data Management 359

- 13.1 Free Format Data Files 359
    - 13.1.1 Reading the File 360
    - 13.1.2 Writing the File 361
- 13.2 BATS Format Data Files 362
    - 13.2.1 Storing and Retrieving Model Definitions 362
- 13.3 Summary of Operations 364
- 13.4 Subset Selection 365
- 13.5 Temporal Aggregation 366
    - 13.5.1 Aggregation and Structural Change 367
- 13.6 Editing and Transformations 368
- 13.7 Summary of Operations 369
- 13.8 Items Covered in This Tutorial 370
- 13.9 Finale 370

Part C: BATS REFERENCE

Chapter 14   Communications 373

- 14.1 Menus 373
- 14.2 Lists 375
- 14.3 Tables 376
- 14.4 Dialogue Boxes 378
- 14.5 Graphs 379

Chapter 15   Menu Descriptions 381

- 15.1 Menu Trees 381
    - 15.1.1 Main Menu Tree 381
    - 15.1.2 Floating Menus 384
- 15.2 Root 385
- 15.3 Data 386
- 15.4 Data/Input-Output 387
- 15.5 Data/Input-Output/Read 389
- 15.6 Data/Input-Output/Write 389
- 15.7 Data/Explore 390
- 15.8 Data/Explore/Seas 393

15.9   Reset   393
15.10 Model   394
15.11 Model/Components   395
15.12 Model/Components/Trend   396
15.13 Model/Components/Seasonal   396
15.14 Model/Interrupts   397
15.15 Model/Discount   397
15.16 Fit   398
15.17 Fit/Existing-Prior   398
15.18 Fit/Existing-Prior/Seasonal   399
15.19 Forecast Priors   400
15.20 Configuration   400
15.21 Forward Intervention   402
15.22 Monitor Signal   403

Index 405

Preface

This book is about practical forecasting and analysis of time series. It addresses the question of how to analyse time series data—how to identify structure, how to explain observed behaviour, how to model those structures and behaviours—and how to use insight gained from the analysis to make informed forecasts.

Examination of real time series motivates concepts such as component decomposition, fundamental model forms such as trends and cycles, and practical modelling requirements such as dealing coherently with routine change and unusual events. The concepts, model forms, and modelling requirements are unified in the framework of the dynamic linear model.

A complete theoretical development of the DLM is presented, with each step along the way illustrated through analysis of real time series. Inference is made within the Bayesian paradigm: quantified subjective judgements derived from selected models applied to time series observations.

An integral part of the book is the BATS computer program. Designed for PC compatible machines, BATS is supplied in both DOS and Windows versions. Completely menu driven, BATS provides all of the modelling facilities discussed and exemplified in the book; indeed, all of the analyses in the book were performed using the program.

There are over 50 data sets in the book. Several are studied as detailed applications, several more are presented with preliminary analyses as starting points for detailed exercises, the remainder are there for further exploration. The data sets are included on the BATS diskette, in readable ASCII format, ready to use with BATS.

The book is organised into three parts. The first part is a course of methods and theory. Time series modelling concepts are described with 'abstract' definitions related to actual time series to give empirical meaning and facilitate understanding. Formal analytical machinery is developed and the methods applied to analysis of data. Three detailed case studies are presented, illustrating the practicalities that arise in time series analysis and forecasting. The second part is a course on applied time series analysis and forecasting. It shows how to build the models and perform the analyses shown in the first part using the BATS application program. Data sets are discussed from a modelling perspective, then the model building procedure in BATS is described and illustrated in the manner of a walk-through tutorial. Steps in the analyses performed are motivated by examination of the nature and structure of the data series, relating ideas back to the discussion and examples in the first part. The third part of the book provides a complete program reference. It is designed to assist in your own modelling once you have been through the tutorials in the second part.

Part A comprises Chapters 1 through 7 and two appendices. Chapter 1 is an introduction to the nature and philosophy of time series analysis and forecasting with dynamic models. Chapter 2 develops, at a non-technical level, the concepts and modelling approach of the dynamic linear model. Mathematical detail is kept to a minimum here, the discussion concentrating on ideas. Chapter 3 parallels Chapter 2 in structure, developing the detailed analytical framework of the univariate DLM. Separated this way, Chapter 2 describes what is being done and why, and Chapter 3 details how it is done. There are two appendices to Chapter 3. Appendix 3.1 provides a summary of distribution theory utilised in the DLM. Appendix 3.2 gives an introduction to classical time series models, autoregressive integrated moving average or ARIMA models and exponentially weighted moving average or EWMA models, showing how these models may be treated as special cases of the dynamic linear model. Chapters 4, 5, and 6 present a series of detailed case studies. The applications are drawn from diverse areas in commerce and society, showing the methods we describe in practice. Chapter 7 contains brief summaries and preliminary analyses of data sets from a wide range of application areas, together with suggestions for investigation.

Part B comprises Chapters 8 through 13. These chapters are designed to be worked through in sequence. Chapter 8 shows you how to set up the BATS program for use on your computer. Chapter 9 introduces you to the program and familiarises you with its operation. Chapter 10 is the start of the real fun! Beginning with the simplest DLM, models of increasing complexity are built and applied to a candy sales series. Chapter 11 demonstrates interactive forecasting; numbers of road traffic related injuries are

Preface

analysed. Chapter 12 illustrates the practicalities of analysing time series that contain missing values. Finally, in preparation for investigating your own time series, Chapter 13 describes how to format data files for use with BATS.

Part C is a two chapter program reference to BATS. When you begin to explore your own time series this is the place to find the commands you need for the analysis that you want to do.

Exercises are included throughout the text. Chapter 3 contains a set of technical exercises designed to explore, enhance, and extend understanding of the formal structure of dynamic linear models. Elsewhere the exercises are tasks of practical time series analysis and forecasting. Questions are posed about the analyses presented in the text, extensions and alternatives, and for analysis of other data sets included on the program diskette. Some of the practical exercises require extensive work and are well suited to term papers and course evaluations. The exercises are an integral part of the text: many subtleties are raised therein that are either not directly addressed, or treated only cursorily, in the text. All the exercises should be read—and preferably most of them attempted.

We have strived to write a text suitable for a broad swathe of undergraduate and beginning graduate students. The applied nature of the book should also appeal to operations researchers and practitioners. Indeed, the book is designed for anyone who deals with time series data and needs to understand and apply appropriate methods of analysis. Data series analysed in the text are drawn from business, economics, environmental studies, geology, medicine, and psychology; data sets from other disciplines including engineering, finance, biology, earth science, speech therapy, travel, agriculture, and astronomy are provided on the accompanying program diskette. Prerequisite for reading the technical material in the text is basic introductory material, as in first courses, in three areas: linear algebra (vector and matrix operations); probability (probability calculus including conditional probability, Bayes' theorem); and statistical distribution theory (multivariate normal distribution, gamma and Student t distributions).

Suggestions for Using this Book

We have organised the material into the two main parts based on our own teaching experiences. We have successfully used the material from parts A and B in parallel. Discussion of concepts and theory from Part A followed by practical illustration from Part B with students involved in hands-on experimentation from the very first class. The menu driven program reduces the learning curve to a minimum, students being able to perform their own analyses in less than an hour. Ideas are introduced sequentially in both parts making the approach quite straightforward. Moreover, the tutorials reinforce many practical points that arise when actually doing analysis; they offer much more than mere lists of instructions on how to perform an analysis with BATS.

The text may also be used in other ways. For those already familiar with dynamic modelling, Part A will function as a brief revision course and concentration will be on the BATS implementation in Part B and the reference in Part C. The organisation also facilitates revision of concepts, theoretical details, or matters of practical implementation.

Acknowledgements

Many people have contributed to the development of the work set out in this volume. We wish to record our sincere appreciation to all of them. In particular, we thank the following people who contributed most visibly. Eliza and Terpsis Michalopoulou researched the Greek marriage data analysed in Chapter 6. Steve Carpenter provided the Lake Mendota data and modelling suggestions in Chapter 7. Neville Davies has the distinction (along with his students and ours) of being chief independent guinea pig during the development of BATS and the material in Part B. Val Johnson reviewed portions of the text in his own inimitable style. Giovanni Parmigiani provided corrections for Chapters 1 and 2. Erik Moledor did the same for Chapter 3. John Pole proof read the entire manuscript. Chris Pole converted BATS to a Windows application. Many users of BATS pointed out bugs and suggested numerous improvements to the program and the material in Part B. Robin Reed made available his TEX macro package for table formatting.

In addition, AP wishes to thank Pete Boyd for many a shared evening and ride home after working late; Cheryl Gobble and Pat Johnson for adminstrative excellence and taking care of him during his time at ISDS; Sue Shelton for being a good friend; Franklin Lee for installing TEX at Blackhawk and helping with Unix utilities; John Singer for keeping the system running. ISDS and Blackhawk provided computer resources.

Finally we would like to record our appreciation of John Kimmel at Chapman and Hall for his patience with very tardy authors!

New York, New York, USA   ANDY POLE
Durham, North Carolina, USA   MIKE WEST
Warwick, UK   JEFF HARRISON
May 1994

Applied Bayesian Forecasting and Time Series Analysis

Part A

DYNAMIC BAYESIAN MODELLING
Theory and Applications

Chapter 1

Practical Modelling and Forecasting

In this introductory chapter we discuss what is special about a time series compared with data that does not have a time component. We examine the nature of time series analysis and forecasting, discuss the importance of dynamic systems, and explore the nature of Bayesian modelling.

1.1 The Nature of Time Series

A time series is a series of observations taken sequentially over time. In a standard regression model the order in which observations are included in the data set is irrelevant: any ordering is equally satisfactory as far as the analysis is concerned. It is the order property that is crucial to time series and that distinguishes time series from non-time-series data. Actions taken at some time have consequences and effects that are experienced at some later time. Time itself, through the mechanism of causality, imparts structure into a time series.

In typical applications successive observations will be equally spaced: daily, weekly, monthly, for example. But such spacing is only a convenience. It is certainly not a necessary restriction. Any sequence of observations can be transformed into an equally spaced series merely by the adoption of a suitably fine time scale. Between adjacent observations separated by more than one unit on this granular scale, missing values are recorded. Bayesian analysis takes missing values in its stride; no special arrangements are needed to handle them.

Extra care must be taken to ensure that the same kind of quantity is being measured when recording intervals are not constant. If adjacent observations are in some way dependent upon the length of the recording interval, then further adjustment is necessary. To see this consider two alternative situations. In the first, sales are recorded weekly, but occasionally no measurement is made. Here the underlying time scale is weekly and the series includes missing values. In the second, sales are recorded weekly as before, but sometimes individual weekly figures are not available, cumulative sales for two or more weeks being recorded instead. Once again the underlying time scale is weekly, but in this case the missing information is of a subtly different nature: the apportioning of cumulated sales across constituent weeks.

In this book attention is devoted entirely to univariate time series wherein a single quantity is the focus of interest, monthly product sales for instance. This does not mean that the models are devoid of explanatory variables. Quite the reverse; such variables will be seen to play an important rôle in many models. By one dimensional we mean the dimension of the response, or dependent variable to use regression theory terminology.

There are, of course, many areas where a univariate linear model is less than optimal, sales of related products for a multiline company for example. Even here, though, univariate models are a useful starting point for more complex multivariate modelling.

1.2 Time Series Analysis vs Forecasting

Forecasting and the analysis of time series are two distinct activities. A forecast is a view of an uncertain future. Time series analysis is a description of what has already happened. One can analyse time series data without regard to forecasting, just as one can make a forecast without regard to time series (or any other) analysis. But our concern is with the forecasting of time series. The main precursors to the forecasting activity are the construction of a suitable model based upon analysis of the historical development of the series and utilisation of information relevant to the series' likely future development.

That is not to say that the modelling and analysis process is concluded once a forecast has been produced. Eventually the quantity being forecast will become known, this value providing new information to incorporate into the analysis. Thereafter, in a typical live situation, a forecast for a subsequent time will be produced and the forecast-observation-analysis cycle repeated.

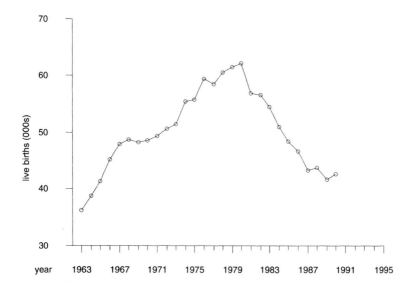

Figure 1.1 Annual live births (000s) in Athens.

1.3 Model Forms

Three basic model forms encompass the great majority of time series and forecasting situations. They are models for simple time trends, systematic cyclical variation, and influential or causal variables (regressions). Combinations of these three basic forms—possibly with a variance law included—provide a large and flexible class of dynamic models suitable for modelling time series in a wide range of diverse applications: business applications such as sales, inventories; industrial applications like production processes, capacity utilisation; agricultural applications like milk yield from cows, meat markets; medical applications in patient monitoring, organ transplants, drug therapy; social statistics such as road traffic accidents; and many more besides. We study applications in several of these areas, detailing data exploration, model building, analysis, and forecasting.

Trend models are the simplest component models in the dynamic linear model toolbag. They represent a system with a straightforward linear progression: growing, decreasing, or staying roughly constant. Figure 1.1 shows the number of live births (in thousands) registered in the city of Athens, Greece, for the years 1963 to 1990. The linear form of this series is self-evident; but notice the very clear change from strong growth to equally strong decline in 1979-80.

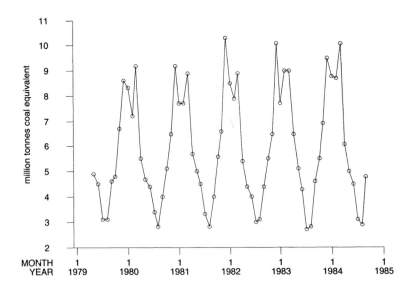

Figure 1.2 Total monthly gas consumption in the U.K.

Seasonal models provide the mechanism to model systematic cyclical variation. This kind of variation is often present in commercial series, and typically related to the passage of the seasons as the earth orbits the sun during the course of a year. Figure 1.2 shows monthly gas consumption in the U.K. from 1979 to 1984. A stable, extremely pronounced, annual cycle is the main feature of this series. Other cycles, with driving mechanisms unrelated to solar progression—biological or mechanical for example—also occur.

Regressions on explanatory variables are potentially the most valuable models because they may incorporate much external information. Figure 1.3 shows logarithmically transformed daily base flow rate and daily precipitation as measured at the San Lorenzo Park gauging station by the U.S. Geological Survey. The series covers the period July 1984 to July 1990. The result of precipitation is clearly discernible in the flow rate series. In the remainder of this chapter we use the example of product sales and prices to illustrate points and concepts.

1.4 Dynamic Systems

Most statistical analysis—theory and practice—is concerned with static models: models with one set of parameters whose values are fixed across

1.4 Dynamic Systems

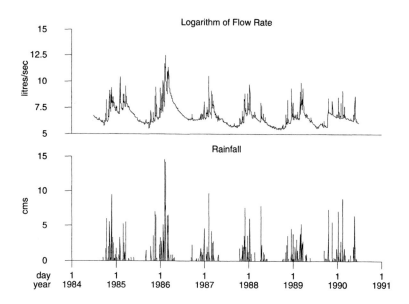

Figure 1.3 Mean daily flow rate (log) and precipitation at San Lorenzo Park.

all experimental or observational units. This is true of regression models. It is also the case in classical time series models, autoregressive integrated moving average models (ARIMA), and exponentially weighted moving averages (EWMA).

Static model formulations implicitly assume that the quantified relationships remain the same across the design space of the data. For example, the response of sales to price is constant whatever the level of sales or prices. In some situations this is a perfectly reasonable approximation and is quite valid. In other situations it is not. In time series it is a very dangerous assumption. The passage of time alone always brings changed circumstances, new situations, fresh considerations. Sometimes entire relationships have to be rethought after the occurrence of some unusual event: the mechanics of surface runoff and river flow following an earthquake, for example. Series directly related to human activity, commercial activities providing many good examples, are notoriously susceptible to changes through time.

Take another look at Figure 1.1. The birth series is strongly linear, but very clearly there is a sharp change from growth to decline at the end of 1979. Less dramatic, but equally important, growth changes are evident around 1967–8 and 1972–3. Indeed, the entire period is characterised by a more or less smoothly changing growth rate.

The models we present in this book are dynamic. They are explicitly formulated to allow for changes in parameter values as time passes. These models contain, as special cases, all of the static models alluded to above. The class of dynamic linear models, or DLMs, is therefore very large and flexible indeed. It is a powerful toolkit for statistical analysis of data.

The passage of time, with additional observations becoming available sequentially, is usually associated with increasing knowledge, thereby adding to the information store. Parameter estimates for static models clearly adjust in such a setting. As additional observations are included in a regression model, parameter estimates do change. So what exactly do we mean by dynamic? Just this. The passing of time induces a loss in the value of information. Because of the many factors that change with time, any information we have at a particular time, say yesterday, will be less directly relevant at another time, say today. When estimating the level of sales for this month, sales figures for six months ago are potentially less useful than sales for last month. This information loss or, equivalently, parametric movement through time, is quite distinct from the mere racking up of a new observation. It is saying: pay more attention to more recent data than to older data.

Why is such temporal adjustment necessary? What advantages does it offer? Answers to these important questions lie in the fundamental modelling process. By definition a model is just a representation of a mechanism, say, a sales series. Unless the mechanism is itself rather simple, and such is seldom the case, a usable model must employ many simplifications and assumptions. It must capture the essence of important relationships without including insignificant minutiae. And therein lies the root of the dynamic 'problem'. The effects of the simplifying assumptions and neglected detail will be different as time changes. Consider, for example, a regressor variable having a quantitatively small contribution to a series and thereby omitted from the current model description. As the variable's value changes with time the neglected effect also changes. Some allowance is made for this in a model having a dynamic parameterisation: the model dynamics allow for adjustment for small errors or omissions that arise from model misspecification. More than this, of course, are direct changes in relationships: a new competitor, a shortfall in the supply of a raw material, altered perceptions caused by rumours or publicity stunts, for example. These changes can be extremely prevalent in any series strongly influenced by human activity, as already mentioned.

When we build a dynamic model we define a qualitative form for the underlying structure of a series. Sales depend on price for example. But any given quantification of the relationship is only locally appropriate. Static models are a special case where relationships apply globally.

1.5 The Bayesian Approach to Forecasting

Model building is an art. There are, of course, guidelines and principles to follow when engaged in the activity, but there are no hard and fast rules by which 'the' model can be derived. This is surely true since there is no single 'true' model. The value of any particular model lies in the efficacy with which it performs the task for which it is constructed. Many models may be built for the same data, and different models will be more or less appropriate for different uses. Even for a single well defined purpose there will usually be a set of competing alternative models that are more or less 'appropriate' or useful for the said purpose. Choosing among these alternative models is something that requires knowledge and experience, both of modelling and of the application area. It is, in the end, a personal, subjective decision. Moreover, as circumstances alter, the choice—even for a fixed purpose—may change. The stock dividend example in Chapter 7 is a case in point.

We have already stated that a forecast is a statement about an uncertain future. It is a belief, not a statement of fact. Representing uncertainty in scientific analysis is the province of probability, its practical application the domain of statistics. Whenever we make a forecast we actually make a statement of probability, or, more generally, state a probability distribution that quantifies the nature of our uncertainty. Any and every forecast is predicated upon a fount of knowledge; forecasts are therefore conditional probability statements, the conditioning being on the existing state of knowledge. If we change our knowledge base, then, typically, our forecast will change.

Knowledge is available in several forms, a useful classification being into historical information and professional wisdom or expertise. From analysis of historical information we may derive a model for our quantity of interest—a forecast model. But other knowledge that is not—cannot be— part of such a formal model may also be valuable. It should not simply be discarded or ignored. Rather it should be brought into play, utilised in combination with the formal model. The Bayesian paradigm provides a rational, coherent, formal framework for combining information: routine model forecasts (or individual model component quantifications, and more generally component forms too) are adjusted by subjective intervention to reflect so called 'external' information. One could say that formalised subjective judgement is the pilot of the Bayesian inferential machine.

Bayesian forecasting proceeds according to what is known as the principle of management by exception. What this means is that during routine periods, routine forecasts from a model will be used directly. But when non-routine events occur, model forecasts will be critically examined by the forecaster and often altered to reflect circumstances that are not included in the model. This synthesis of the routine model output with external informed opinion is the centrepiece of Bayesian inference.

Take the example of sales forecasting for a product in a competitive market. While the market is in equilibrium a given firm will forecast its future monthly sales using its routine forecasting model. What happens when a market disequilibrium occurs? Suppose that the manufacturing plant of a competitor is destroyed. There is no possibility that such an occurrence will be part of our firm's routine forecasting model, and therefore no way that the model can reflect the new market conditions. The Bayesian forecaster, however, is in precisely the position required to make such adjustments. Existing knowledge of the competitor's market share and other pertinent factors, such as availability of alternative production facilities, will enable our forecaster to form a judgement about the likely effect of the competitor's misfortune on his own firm's sales. The Bayesian forecaster needs only to cast this judgement—and it is certainly a judgement, not an observed fact or fixed item—in terms of probability statements about the quantity of interest, the forecast series. A new forecast is then made, one that combines the routine model information with the unusual event implications.

The Bayesian paradigm, as already remarked, is logical, consistent, and coherent. But it is more than a set of formal rules to be blindly followed; the illustration above should make that clear. Bayesian thinking is disciplined common sense (Smith, 1991). Common sense is desirable because it is understandable, a comfortable position to hold. But it can be misleading, most especially where uncertainty is involved. Nature abhors a vacuum and the human mind abhors uncertainty. It is not professional expediency that makes probability theory arcane and sometimes counterintuitive. It is the nature of uncertainty. The discipline of Bayesian thinking and the formal learning mechanism of Bayes' theorem avoids logical problems by guaranteeing the validity of conclusions following from given premises. We can be comfortable with Bayesian thinking. We can rely on the integrity of conclusions derived from its application.

1.6 Future Developments

We concentrate entirely on the class of univariate dynamic linear models in this book. Extensions to multiprocess models, multivariate models, and non-normal and nonlinear models are beginning to find increasing application. At the present time, however, experience is limited and no general purpose computational tools exist. Work on all these areas—theory, practice, and implementation—is currently underway.

1.7 Reference Material

The book by West and Harrison (1989) to which we refer on several occasions in this text is the most complete reference currently available for dynamic Bayesian modelling. That volume is strongly recommended as a reference text, containing a complete mathematical description of the structure of dynamic models as it presently exists, and many references to the literature. It includes numerous theoretical exercises making the text particularly suitable for instruction at an advanced level. Readers who progress beyond the present work will find much of interest there...and may even be stimulated to pursue research of their own in this field.

Box, G.E.P. and Tiao, G.C. (1973). *Bayesian Inference in Statistical Analysis*. Addison-Wesley, Massachusetts.

Smith, A.F.M. (1991) Conference discussion, Carnegie Mellon University.

West, M. and Harrison, P.J. (1989). *Bayesian Forecasting and Dynamic Models*. Springer-Verlag, New York.

Chapter 2

Methodological Framework

We present in this chapter an overview of the ideas and concepts of the Bayesian analysis of the dynamic linear model. The treatment is discursive but not overly technical, designed to provide a hook into the mathematical and analytical structure without obscuring the focus with detail. With this groundwork in place, the individual elements introduced and elaborated in the modelling discussion will fit recognisably into the overall framework. An extensive formal development is given in Chapter 3.

2.1 The Dynamic Linear Model

We begin with the general linear model with which you should already be familiar. For a vector of dependent, or response, variables Y, and matrix of independent, or regressor, variables X, the linear model is usually written

$$Y = X\theta + \nu.$$

Here, θ is a vector of unknown parameters and ν is a vector of uncorrelated, unobservable stochastic error terms.

In the dynamic, or time varying, context the linear model is expressed in a form similar to the standard form above. The equation for a single observation is

$$Y_t = x_t'\theta_t + \nu_t,$$

where x is a column vector and x' is its transpose. Notice that apart from the time subscript on the parameter vector this equation is precisely of the form of a single row of the standard equation. Immediately the dynamic model exhibits two generalisations over the standard model—referred to as the static or time invariant model in a time series context. First, there is the explicit time ordering of the observation sequence, Y_1, Y_2, ..., Y_t. Second, there is an individual parameterisation for each time, hence the time subscripted parameter vector, θ_t.

It may appear such a formulation cannot be practically useful since it contains more parameters than observations. With only a single observation from which to derive information on the parameter vector, estimation will be impossible except for the trivial single parameter model. Even then there will be a one-to-one correspondence between parameter values and observations and so no possible measure of uncertainty. Further information is clearly required.

The parameter sets for each observation in the linear model are related—they are the same set. Every observation contains information on each parameter. By contrast, in the dynamic linear model, parameter sets are distinct, but are stochastically related through the *system equation*. The system equation describes the model dynamic, the mechanism of parametric evolution through time. This evolution has the general form of a first order Markov process,

$$\theta_t = G_t \theta_{t-1} + \omega_t,$$

where G_t is a matrix of known coefficients and ω_t is an unobservable stochastic term. The system evolution matrix G_t defines a known deterministic functional relationship of the state vector (parameter vector) at one time with its value at the immediately preceding time. Through the system equation information on the state vector is propagated through time. But the presence of the stochastic element, ω_t, allows for an imperfect propagation, a fuzziness or information loss. This models precisely what was discussed earlier: the passage of time erodes the value of information—sales figures for six months ago are potentially less useful than figures for last month when predicting sales for this month.

To put the model details in a concrete setting consider once again the sales-price illustration. Sales is composed of two parts: an underlying level plus an amount determined by price. The observation equation for the model may be expressed as

$$\text{sales}_t = \text{level}_t + \beta_t \text{price}_t + \nu_t, \qquad \nu_t \sim N[0, V].$$

The notation "$\sim N[0, V]$" represents "has a normal distribution with mean zero and variance V". Assuming, for brevity, that the observation variance, V, is known, there are two uncertain parameters in this equation: the current level and the current price regression coefficient. The system equation

2.1 The Dynamic Linear Model

for both of these parameters has a particularly simple form, a simple random walk

$$\text{level}_t = \text{level}_{t-1} + \Delta\text{level}_t, \qquad \Delta\text{level}_t \sim N[0, W_1],$$
$$\beta_t = \beta_{t-1} + \Delta\beta_t, \qquad \Delta\beta_t \sim N[0, W_2].$$

The notation "Δx" denotes an increment in the quantity x. As you can see, this model says that the level today is expected to be the same as it was yesterday, but there is some chance that it might be different. The amount of movement in the level over time is controlled by the stochastic term, Δlevel_t, or more specifically, the variance, W_1, thereof. As this variance increases, the more volatile the series becomes. Conversely, as variance decreases, the less volatile the series becomes, and at the limit $W_1 = 0$ there is no volatility, the dynamic model reduces to the familiar static model: the level does not change with the passage of time.

The comments just made with respect to the level parameter also apply to the regression coefficient since the structure of the level and price coefficient evolution is identical. For more general models the same ideas of deterministic movement and information decay apply to every element of the state parameter vector albeit with possibly more complex evolutions.

Before we begin the analysis of the DLM let us place the observation equation and two system equations above in the standard notation. This notation is used throughout the book and in West and Harrison (1989). Define $\theta = (\text{level}, \beta)$, $F' = (1, \text{price})$, and $Y = \text{sales}$. Then the DLM is written

$$\text{Observation Equation}: \quad Y_t = F_t'\theta_t + \nu_t, \qquad \nu_t \sim N[0, V],$$
$$\text{System Equation}: \quad \theta_t = G_t\theta_{t-1} + \omega_t, \qquad \omega_t \sim N[0, W].$$

For the sales example the system matrix is the identity, $G_t = I$, and the system variance matrix is the diagonal form $W = \text{diag}(W_1, W_2)$.

2.1.1 Time and Order

We emphasised above that the explicit time ordering of the observation sequence is an important piece of the structure of dynamic models. This is certainly true in a time series context but dynamic models are more widely applicable than 'time ordering' would suggest. Any data series for which there is a natural ordering of observations fits into the dynamic framework. Consider spatial data for example. Points in space, areas of agricultural land, sea floor surface areas, and so on all have natural order properties that relate directly to underlying structure. Adjacent areas are more likely to be structurally similar than areas separated by some distance. Dynamic modelling is as appropriate for studying spatially ordered series as it is for time ordered series. It is also appropriate for series with any other ordering criteria. The ordering itself imparts a structure to data series; it is this structure—not the underlying characteristic such as time or space—that dynamic modelling is designed explicitly to address.

2.2 Bayesian Analysis

Bayesian statistical analysis for a selected model formulation begins by first quantifying the investigator's existing state of knowledge, beliefs, and assumptions. These prior inputs are then combined with the information from observed data quantified probabilistically through the likelihood function—the joint probability of the data under the stated model assumptions. The mechanism of prior and likelihood combination is Bayes' theorem, a simple theorem of inverse probability. The resulting synthesis of prior and likelihood information is the posterior distribution or information. In technical terms (Chapter 3) the posterior is proportional to the prior and the likelihood,

$$\text{posterior} \propto \text{prior} \times \text{likelihood}.$$

This prior to posterior process is referred to as Bayesian learning. It is the formal mechanism through which uncertainties are modified in the light of new information.

2.2.1 Sequential Analysis

The sequential nature of time series, with data becoming available in temporally distinct parcels, means that the learning procedure is also sequential. The prior analysis just outlined is performed every time a new observation is made.

Bayesian learning takes us along the path from today's starting position—before today's sales are seen—to today's final position—after today's sales are seen, updating what we knew this morning with what we observed during the day. The next step is to progress from today to tomorrow. Believing what we do about the level of sales today, what will we think about the level for tomorrow when we wake up in the morning? Given the posterior information for today, what does this imply about the prior information for tomorrow? You may have already determined the answer from the discussion on temporal linkage in the previous section. The system evolution equation specifies precisely how posterior beliefs about today's level relate

2.4 Forecasting

to prior beliefs about tomorrow's level. In probabilistic notation, after observing today's sales we have posterior beliefs on the underlying sales level summarised as a distribution

$$p(\text{level}_t | D_t).$$

(More generally our beliefs about the whole state vector are summarised as a joint distribution, $p(\theta_t | D_t)$.) The posterior distribution for the level quantifies our relative degree of belief in the possible values of the level conditional on all the pertinent information up to and including today, D_t. Moving to tomorrow, before we receive tomorrow's sales figures we state our beliefs as the prior distribution. For the level we express this as

$$p(\text{level}_{t+1} | D_t),$$

which is derived from the previous posterior using the system evolution equation (see Chapter 3 for details). Notice the difference in the time subscripts here. The prior relates to a view of tomorrow given information only up to today. It is a forward looking view. The previous posterior refers to a view of today given that same information: a rearward looking view. The prior is more uncertain than the previous posterior by an amount determined by the evolution variance W_t. As we have said, tomorrow is a new day and things might change. We are less sure about what is yet to be than about what is already history; less sure in our view of what tomorrow's sales level will be than of what today's sales level was.

2.3 Subjective Intervention

In routine operation the sequential development of prior to posterior to prior is the full extent of the analysis. But we emphasised in Chapter 1 that incorporation of external information—we used the example of a catastrophe befalling a competitor—is a major advantage of the Bayesian methodology. How is such external information included?

The prior specification is adjusted directly. For example, if our analysis of the market situation leads us to believe that our competitor's ill luck will result in a doubling of our own sales, we simply reset the mean value of the prior distribution accordingly. Notationally we may write this formally as moving from the pure model based prior $p(\text{level}_t | D_{t-1})$ to the post-intervention prior $p(\text{level}_t | D_{t-1}, I_t)$ where I_t denotes any relevant external information available to us at the time. The change in the prior distribution reflects our subjective judgement of the effect on sales of the additional information (and typically includes adjustment of the variance as well as the mean).

2.4 Forecasting

Model forecasts are derived from the prior information and the observation equation. For our sales model the observation equation is

$$\text{sales}_t = \text{level}_t + \beta_t \text{price}_t + \nu_t, \qquad \nu_t \sim N[0, V].$$

Given prior information on the level and price regression coefficient, and the error variance V, we simply combine the probability distributions using the (linear) observation equation. The resulting predictive distribution for today's sales given information up to yesterday is

$$p(\text{sales}_t | D_{t-1}).$$

Once again, the general theoretical development is detailed in Chapter 3.

Forecasting several periods into the future is done in an analogous manner. For example, a sales forecast for two periods hence (once again from the standpoint of the beginning of today, so our information includes outcomes only up to yesterday) requires the distribution

$$p(\text{sales}_{t+2} | D_{t-1}).$$

This distribution is obtained by repeatedly applying the system evolution equation to the expression for sales_{t+2} to obtain an expression involving the quantities about which we have information—namely the priors on today's level and price coefficient:

$$\begin{aligned}
\text{sales}_{t+2} &= \text{level}_{t+2} + \beta_{t+2}\text{price}_{t+2} + \nu_{t+2} \\
&= (\text{level}_{t+1} + \omega_{1,t+2}) + (\beta_{t+1} + \omega_{2,t+2})\text{price}_{t+2} + \nu_{t+2} \\
&= (\text{level}_t + \omega_{1,t+1} + \omega_{1,t+2}) \\
&\quad + (\beta_t + \omega_{2,t+1} + \omega_{2,t+2})\text{price}_{t+2} + \nu_{t+2},
\end{aligned}$$

where we have used ω_1 to denote Δlevel and ω_2 to denote $\Delta\beta$. This rolling back procedure is exactly equivalent to formally propagating the prior information we have about today to the implications it has for two periods hence in the absence of any additional pertinent information.

In a more general situation we may be aware of significant changes that will affect future sales. For example, we might know that our competitor will be able to restart production in two period's time. This is expected to lead to a loss of the temporarily increased demand currently being experienced, and the forecast would be modified as described in the previous Section.

It is often desirable to compute a cumulative forecast for several periods ahead. Manufacturing offers a good example. When ordering parts or raw materials that are most economically obtained in large batches, orders are placed in advance for several weeks worth of production. The preceding paragraphs demonstrate how to calculate one period forecasts for any number of steps into the future; cumulative forecasts for a number of periods ahead simply require an appropriate addition of these individual forecasts. A little care is needed though. When adding quantities described by probability distributions attention must be paid to covariances among terms. Following our previous example, calculate the cumulative forecast for the next two time periods,

$$\text{sales}_{t+2} + \text{sales}_{t+1}.$$

The previous development shows how each of the terms in the sum can be rolled back into sums of quantities for which we currently have full probabilistic descriptions. All we then need do is add both sets of terms together. Distributional expectations simply add. But some of the stochastic terms now appear more than once. The variances of such terms contribute to the variance of the sum in an amount equal to the *square* of the number of times they appear. If a term appears twice, its contribution to the overall sum is four times its unary value. Covariances between different stochastic terms must also be included; however, the conventional assumptions of the DLM make these terms independent, so all such covariances are zero and can therefore be ignored.

2.5 Distributional Forms

The discussion of the DLM thus far has been carried out in terms of general probability distributions $p(\cdot)$ and latterly has concentrated on the first two central moments—the expectation (mean) and the variance. Also, if you examine the foregoing analysis you will note that stochastic terms appear only in linear combinations. Now, the standard assumptions for the DLM are that all stochastic terms are normally distributed, and basic theory shows that linear combinations of normal variables are also normally distributed (see Appendix 3.1). Therefore, all the defining distributions of the DLM are normal. Since the mean and variance completely characterise the normal distribution we need add nothing further to the analysis. Chapter 3 details distributional forms in full and presents the formulae which are used to update moment estimates at each time.

Concentrating on moments rather than full distributional forms the analysis of the DLM outlined above applies much more generally than for just normal models. Under 'weak probability modelling' the derived means and variances from the normal distribution assumptions may be used with other distributional forms. In a decision-theoretic setting where decision rules are functions of moments (regardless of underlying distributions), such as in linear Bayes analysis (West and Harrison, 1989), the moment definitions from the normal analysis are again sufficient.

2.6 Monitoring

Model selection does not end with the production of a forecast. Eventually the quantity being forecast will be measured in outcome and it will be possible to gauge the accuracy of the forecast. Competing models may be judged on the basis of their forecast performance. The relative suitability of alternative models at different times provides information about the implications of unexpected or unusual events.

The raw material of a forecast is a probability distribution. For the normal probability distribution any observed value is theoretically possible. (This is an idealisation of course, and it is seldom truly appropriate. Sales of an item cannot be negative—and typically not fractional—for instance.) Observations occurring in the tail of the distribution are always possible—by definition since every value has nonzero probability—but unlikely. And the further out in the tail of the distribution the more unlikely the observation.

Unlikely events are occasions of which we would like to be made aware. If an unlikely event occurs, it may be the result of unusual circumstances. If so, those circumstances ought to be examined. Are they transient? Or will they have implications for the future? Perhaps adjustments should be made to the model before further forecasts are produced?

Observations that have a small chance of occurring under an assumed model—values in the tail of the forecast distribution—can be automatically flagged by the forecast system. Specify a rule which encapsulates the nature of 'unusual'—simply a threshold on observation-forecast inconsistency—then issue a signal when the threshold is breached.

Making a judgement of forecast-outcome consistency for a single model is only part of the assessment story. Stating that an observation is unlikely (to some specific degree) with a given model only points to potential general deficiency of that model. More information is necessary to identify and correct any deficiency. It would be helpful to know in what sense the model is deficient. Better yet, what alternative model would have outperformed the routine model?

A forecast distribution provides a numerical description of the relative degree of belief in the possible values that a quantity may exhibit. Alternative models are contrasted by the differences in relative degrees of belief of possible outcomes. Each set of beliefs characterises a different forecast distribution and therefore has a different measure of consistency with the observed value. These measures of consistency help identification and assessment of possible model deficiencies.

What are these alternative sets of beliefs, these alternative models? Each departure from the routine model defines an alternative model: an increase in level, a change in volatility, a new market structure for example. Ranking these models in terms of their relative forecast-outcome consistency provides a view of whether an observation is unusual, and if so, what makes it unusual.

An automatic monitoring scheme can be set in place to watch over the quality of model forecasts. When the routine model is outperformed by an alternative a signal may be generated to alert the forecaster to a potential problem. Not all signals will be portents of real problems, of course. Occasionally, observations do occur that have small probability of occurring with the routine model, even though there is really nothing untoward occurring in the series. Signal response must therefore be a judgement call. One must look to the system under investigation to determine if there is reason for possible change.

This is how management by exception works. Forecasts are generated from the current routine model, possibly with external information incorporated through prior intervention. When an observation is made, the consistency of forecast and outcome for the routine model is compared with the consistency for alternative models, these alternatives being designed to capture the range of departures from the norm that are of interest. If any of the alternative models is significantly more consistent than the routine model, an exception is signalled. Investigation will provide the forecaster with additional information which may or may not point to adjustment of the routine model before continuing. If change is deemed appropriate, subjective intervention is made at that time. The system then rolls forward to the next time and the cycle of forecast-observation-performance monitoring is repeated.

2.7 Variance Analysis

Implicit in the preceding analysis is knowledge of the observation variance, V, and the system evolution covariances, W. In practice both of these quantities will be unknown and difficult to specify. The Bayesian approach to uncertainty, as already seen for the system state, is to (i) specify information as a probability distribution for the quantity of interest, and (ii) update this distribution (using Bayes' theorem) with observational information (the likelihood) when it becomes available. There is no distinction drawn between parameters measuring a state or a variance. The analysis is straightforward and tractable. Theoretical details for formal learning about the observation variance are given in Chapter 3.

The system evolution covariance matrix can formally be handled in the same general manner: specify a prior distribution and use Bayes learning. However, this becomes exceedingly difficult from an analytical standpoint and practically impossible to implement on a routine basis. Fortunately there is a practicable solution to the problem that very neatly captures both the essence of the evolution mechanism and the spirit of Bayesian thinking. The concept is that of information discounting. The result is a simple, tractable operating procedure.

In our earlier discussion of the nature and value of information on a temporal scale we noted that as information ages it becomes potentially less useful: in some sense its value diminishes. The ageing process is modelled in the DLM through the system evolution. This stochastic mechanism serves to increase uncertainty with the passage of time. The linear structure is additive—some extra variance is added to the state posterior distribution to yield the prior for the next period. In our sales example the prior variance on today's level is, in the absence of further knowledge, somewhat greater than the posterior variance on yesterday's level, this increase representing the information loss attributed to the advancement of time,

$$V[\text{level}_t|D_{t-1}] = V[\text{level}_{t-1}|D_{t-1}] + W_t.$$

Perhaps a more intuitive way of thinking about information loss is in percentage or proportional terms. For example, we might quantify the decay process as a 5% increase in uncertainty as information ages by one time period. In our sales level illustration we would write this as

$$V[\text{level}_t|D_{t-1}] = (1 + \lambda)V[\text{level}_{t-1}|D_{t-1}],$$

with $\lambda = 0.05$. Formal equivalence of the proportional way of writing the variance increase with the standard additive way is seen by setting $W_t = \lambda V[\text{level}_{t-1}|D_{t-1}]$.

Information is usually measured in terms of precision, the inverse of variance: an infinite variance representing no information, and zero variance representing complete information or no uncertainty. Working in terms of information, or precision, we would express the previous relationship in the alternative form

$$\begin{aligned} I(\text{level}_t|D_{t-1}) &= V^{-1}(\text{level}_t|D_{t-1}) \\ &= (1+\lambda)^{-1}V^{-1}(\text{level}_{t-1}|D_{t-1}) \\ &= (1+\lambda)^{-1}I(\text{level}_{t-1}|D_{t-1}). \end{aligned}$$

On this information scale the *discount factor* $(1+\lambda)^{-1}$ varies between 0 and 1 and this is often how we specify discounts in practice. For a discount factor $\delta \in (0,1]$ the information loss through the evolution process is summarised as

$$V[\text{level}_t | D_{t-1}] = \delta^{-1} V[\text{level}_{t-1} | D_{t-1}].$$

For a 5% information loss, $\delta \approx 0.95$.

The discounting procedure applied to information is a natural way to think about the ageing effect, akin to the way discounting is used in economics and commerce. Moreover, discount factors are independent of the measurement scale of the observation series, a significant simplification: a 5% decay is a 5% decay on any scale.

We have begged a question in this presentation of information discounting. How do we choose a suitable discount? Instead of specifying a covariance matrix for the system evolution equation we now have to specify a discount factor. But we are not learning about the discount as we are with each of the other model parameters. It is certainly possible to do so; the model defines a proper likelihood, we can define a prior, and then proceed with Bayes' theorem to obtain posterior inferences. Once again, however, the analysis is not tractable. And the whole point of discounting is that such a full scale analytical approach is not necessary.

In practice we proceed informally by comparing the forecasting performance of a range of possible discount values. The range is not huge for two reasons. First, useful models have discounts not much less than 0.8. Smaller discounts lead to models that make predictions based on only two or three observations regardless of the amount of historical information available. Resulting forecast variances will typically be large and the models of little real value. Second, over the useful range of discounts only a few values need be examined because the likelihood is typically rather flat. The illustrations in later chapters demonstrate both points and give practical guidance on choosing discount factors.

2.8 Smoothing

Time series analysis takes a view of the development of a series using the benefit of hindsight. At the end of a week, looking back one has a clearer picture of what actually happened during that week than one did on any individual day. Later observations contain information about the earlier days and, therefore, uncertainties are reduced. We know more. Such backwards evaluation is variously called smoothing, filtering, or retrospective analysis.

In the sequential analysis outlined above we described the on-line analysis of a series: modifying beliefs sequentially as each new observation becomes available, describing beliefs as probability statements such as $p(\text{level}_t|D_t)$. In the smoothing context these distributions are revised in the light of later information giving filtered distributions $p(\text{level}_t|D_{t+k})$ for some positive k. Notice that the information set D_{t+k} on which the conditioning, or view, is taken includes observations after the time for which the level is estimated.

The mechanism of information filtering is Bayes' theorem. A sequential back filtering procedure is defined analogous to the forward, on-line, development previously described (details in Chapter 3).

2.9 Component Models

The structure of the DLM offers several advantages to the time series modeller. We have already noted the analytical elegance and simplicity that flows from linear combinations of normally distributed quantities. A related and extremely important feature is the component structuring possible with these models. (The structuring ideas we now discuss extend to certain nonlinear models given the basic dynamic model framework.)

On first acquaintance many time series exhibit what appears to be very complex behaviour. A moment's reflection often changes this view somewhat as basic distinctive patterns become apparent. Take a look at Figure 2.1. The initial impression of this series is one of considerable complexity—there seems to be a great deal going on. Now apply the analysis maxim: simplify and conquer. Look beyond the individual observations; forget the time to time changes and look at the overall development in time. Smooth the pattern of the data in your mind and you discern an overall trend. On average the series is increasing at a roughly constant rate for the first seven or eight years, and thereafter it is declining—perhaps a degree more erratically than the increase. Now examine the movement of the series about this perceived trend. There is clearly a systematic cycling about the trend.

This brief inspection analysis has identified the fundamental structure of the series as comprised of two parts: an overall trend in level, and a periodic movement about the trend. The DLM facilitates the direct representation of these component parts as individual sub-models. We build separate component models for the trend and periodic (or seasonal) patterns, then combine the two in an overall model. The resulting model is seen to be structured directly in terms of observable features in the time series so that communication of information to and from the model is in terms of meaningful quantities.

2.9 Component Models

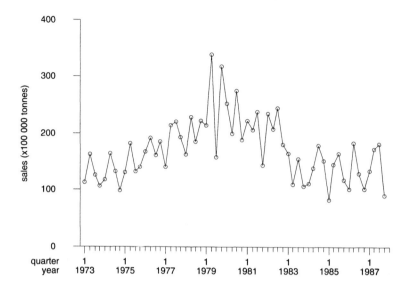

Figure 2.1 Industrial sales series.

Return to our sales example for a moment. We defined sales to have an underlying level and an effect attributed to changes in price. These model component forms have a directly meaningful interpretation.

2.9.1 Block Structuring

The formal algebraic representation of component-wise structuring is quite straightforward. For an observation series comprising two components, a trend and a seasonal periodic component, we can write the observation equation as a sum

$$Y_t = Y_{Lt} + Y_{St} + \nu_t.$$

We construct the appropriate DLM for each of the level, Y_L, and seasonal, Y_S, components so that

$$Y_{Lt} = F'_{Lt}\theta_{Lt},$$
$$\theta_{Lt} = G_{Lt}\theta_{Lt-1} + \omega_{Lt},$$

$$Y_{St} = F'_{St}\theta_{St},$$
$$\theta_{St} = G_{St}\theta_{St-1} + \omega_{St}.$$

The observation equation is a linear combination of these components,

$$Y_t = F'_{Lt}\theta_{Lt} + F'_{St}\theta_{St} + \nu_t$$
$$= F'_t \theta_t + \nu_t,$$

where the regression and state vectors simply stack: $F'_t = (F'_{Lt}, F'_{St})$, and $\theta'_t = (\theta'_{Lt}, \theta'_{St})$. Similarly the system equation is composed of the individual system equations,

$$\theta_t = G_t \theta_{t-1} + \omega_t, \qquad \omega_t \sim N[O, W],$$

where $\omega'_t = (\omega'_{Lt}, \omega'_{St})$, and the evolution matrix and variance have the block structured forms

$$G_t = \begin{pmatrix} G_{Lt} & 0 \\ 0 & G_{St} \end{pmatrix}, \qquad W_t = \begin{pmatrix} W_{Lt} & 0 \\ 0 & W_{St} \end{pmatrix}.$$

Larger models that contain more than two components are built in the same way. Each component contributes a term to the observation equation and a block of parameters to the system equation. Details of the regression vector and system matrix of DLM component models for trends, seasonal patterns, and regressions, are given in Chapter 3.

2.10 Summary

The ideas presented in this chapter, sequential analysis, intervention, forecasting, monitoring, variance discounting, retrospective views, component forms, and block structuring are illustrated and discussed in detail throughout the remainder of the book. Chapter 3 details the formal mathematical and statistical development of all these ideas, and Chapters 4 through 7 discuss several applications. Part B offers a guided tour through analysis of time series data, detailing and illustrating each stage in the argument, and relating the procedures to the concepts in this chapter. Step-by-step instructions are provided for reproducing the analyses with BATS. At this stage you might like to skip to Part B, introduce yourself to BATS and work through the first two tutorials. You will see many of the concepts in action which may help hasten and deepen your appreciation of the technical discussion to come in Chapter 3.

2.11 References

Berger, J.O. (1985). *Statistical Decision Theory and Bayesian Analysis.* Springer-Verlag, New York.

DeGroot M.H. (1970). *Optimal Statistical Decisions.* McGraw-Hill, New York.

West, M. and Harrison, P.J. (1989). *Bayesian Forecasting and Dynamic Models.* Springer-Verlag, New York.

Chapter 3

Analysis of the DLM

Chapter 2 gave a heuristic introduction to the dynamic linear model concentrating on discussion of the conceptual framework omitting technical detail. This chapter fills in that detail. The complete analytical structure of the univariate DLM is presented, with the caveat that some of the more technically complicated results are presented without proof. Where such proofs are omitted references are given. The ordering of the material mirrors the development in Chapter 2 so that both chapters can be easily cross-referenced.

You may find it a useful pedagogical device to work through the tutorials in Part B as you read this Chapter. Those tutorials are designed to exemplify in an applied setting the concepts introduced in Chapter 2 and detailed theoretically here.

3.1 Model Form and Notation

The general univariate dynamic linear model is written as

Observation Equation : $Y_t = F'_t \theta_t + \nu_t, \qquad \nu_t \sim N[0, V_t]$,

System Equation : $\theta_t = G_t \theta_{t-1} + \omega_t, \qquad \omega_t \sim N[0, W_t]$,

where Y_t denotes the observation series at time t; F_t is a vector of known constants (the regression vector); θ_t denotes the vector of model state parameters; ν_t is a stochastic error term having a normal distribution with zero mean and variance V_t; G_t is a matrix of known coefficients that defines the systematic evolution of the state vector across time; and ω_t is a stochastic error term having a normal distribution with zero mean and covariance matrix W_t. The two stochastic series $\{\nu_t\}$ and $\{\omega_t\}$ are assumed to be temporally independent and mutually independent; that is, the covariances $\mathrm{Cov}[\nu_t, \nu_s]$, $\mathrm{Cov}[\omega_t, \omega_s]$ for all $t \neq s$, and $\mathrm{Cov}[\nu_t, \omega_s]$ for all t, s are zero. These assumptions are made for convenience rather than necessity since models with correlated stochastic terms can always be transformed into the setup used here. We use this independence form because the analysis is thereby most easily derived and presented.

In this book we restrict attention to DLMs that have an invariant system matrix, $G_t \equiv G$. Generalisation to time varying evolution forms is straightforward, requiring more complex algebra only. The interested reader may refer to West and Harrison (1989) for details of the analytical extensions.

3.2 Updating: Prior to Posterior Analysis

Bayesian learning proceeds by combining information from observations expressed through the likelihood function with the investigator's existing state of knowledge before the observations are made. The mechanism of combination is Bayes' theorem, a simple theorem of inverse probability.

3.2.1 Bayes' Theorem

For two quantities A and B for which probabilistic beliefs are given, Bayes' theorem states

$$p(A|B) = \frac{p(B|A)p(A)}{p(B)},$$

where the notation $p(x|y)$ denotes a probability density function for the quantity x conditional upon information y. This result enables us to make statements about (say) a level following an observation given (i) a quantification of what we believed prior to making the observation, and (ii) a model for the system generating the observation series. We will see this in the derivation of posterior information below.

3.2.2 Prior Information

Prior information on the state vector for time $(t+1)$ is summarised as a normal distribution with mean a_{t+1} and covariance R_{t+1},

$$\theta_{t+1}|D_t \sim N[a_{t+1}, R_{t+1}],$$

where D_t denotes the state of knowledge at time t.

3.2 Updating: Prior to Posterior Analysis

3.2.3 Forecasts

From the prior information, forecasts are generated using the observation equation.

Forecasting One Step Ahead

The forecast quantity Y_{t+1} is a linear combination of normally distributed variables, $\theta_{t+1}|D_t$ and ν_{t+1}, and is therefore also normally distributed. The forecast mean and variance are easily seen to be

$$\begin{aligned} \mathrm{E}[Y_{t+1}|D_t] &= \mathrm{E}[F'_t\theta_{t+1} + \nu_{t+1}|D_t] \\ &= \mathrm{E}[F'_t\theta_{t+1}|D_t] + \mathrm{E}[\nu_{t+1}|D_t] \\ &= F'_t\mathrm{E}[\theta_{t+1}|D_t] + \mathrm{E}[\nu_{t+1}] \\ &= F'_{t+1}a_{t+1} \\ &= f_{t+1}, \end{aligned}$$

$$\begin{aligned} \mathrm{V}[Y_{t+1}|D_t] &= \mathrm{V}[F'_{t+1}\theta_{t+1} + \nu_{t+1}|D_t] \\ &= \mathrm{V}[F'_{t+1}\theta_{t+1}|D_t] + \mathrm{V}[\nu_{t+1}|D_t] \\ &= F'_{t+1}\mathrm{V}[\theta_{t+1}|D_t]F_{t+1} + \mathrm{V}[\nu_{t+1}] \\ &= F'_{t+1}R_{t+1}F_{t+1} + V_{t+1} \\ &= Q_{t+1}. \end{aligned}$$

The forecast distribution for one step ahead therefore has the normal form $Y_{t+1}|D_t \sim N[f_{t+1}, Q_{t+1}]$. (Note that we used the assumption that the observation disturbance term, ν_{t+1}, is uncorrelated with the state, θ_{t+1}.)

Forecasting k Steps Ahead

Forecasting several steps ahead requires the prior information to be projected into the future through repeated application of the system equation. Given the prior for time $(t+1)$ the implied prior for time $(t+2)$ from the same standpoint, with no additional information, is $p(\theta_{t+2}|D_t)$. This prior is obtained by applying the system equation as in

$$\theta_{t+2} = G\theta_{t+1} + \omega_{t+2}, \qquad \omega_{t+2} \sim N[0, W_{t+2}].$$

Linearity once again ensures that this two step ahead prior will be normal. Defining moments are identified as

$$\begin{aligned} \mathrm{E}[\theta_{t+2}|D_t] &= G\mathrm{E}[\theta_{t+1}|D_t] + \mathrm{E}[\omega_{t+2}] \\ &= Ga_{t+1}, \end{aligned}$$

$$\begin{aligned} \mathrm{V}[\theta_{t+2}|D_t] &= G\mathrm{V}[\theta_{t+1}|D_t]G' + W_{t+2} \\ &= GR_{t+1}G' + W_{t+2}. \end{aligned}$$

By extending this rolling forward procedure it is easy to see that the k step ahead forecast distribution has the normal form

$$\theta_{t+k}|D_t \sim N[a_t(k), R_t(k)],$$

where, for $k \geq 2$, the mean and variance are given by

$$a_t(k) = G^{k-1}a_{t+1},$$

$$R_t(k) = G^{k-1}R_{t+1}(G^{k-1})' + \sum_{j=2}^{k} G^{k-j}W_{t+j}(G^{k-j})'.$$

Note that the sum runs from $j = 2$ because the stochastic evolution variance W_{t+1} is already included in the state prior variance R_{t+1}. Given this forecast for the state the associated forecast for the observation series is obtained from the observation equation as

$$Y_{t+k}|D_t \sim N[f_t(k), Q_t(k)],$$

where the moments are defined in familiar terms,

$$f_t(k) = F'_{t+k}a_t(k),$$
$$Q_t(k) = F'_{t+k}R_t(k)F_{t+k} + V_{t+k}.$$

Cumulative Forecasts

Cumulative forecasts for several periods are easily obtained from the step ahead forecasts just derived. Once again the desired quantity is a sum of normally distributed components so it is necessary only to determine the cumulative forecast mean and variance, the distribution being normal. The mean value is just the sum of the individual forecast means of each period in the cumulation. The cumulative forecast variance is more complicated because covariances among the individual step ahead forecasts—which are not independent—must be determined.

Let Z_{t+k} denote the sum of the values of the observation series over the next k periods,

$$Z_{t+k} = \sum_{j=1}^{k} Y_{t+j}.$$

Then the expected value for the sum conditional on information up to and including today is

$$E[Z_{t+k}|D_t] = \sum_{j=1}^{k} E[Y_{t+j}|D_t]$$

$$= \sum_{j=1}^{k} f_t(j).$$

3.2 Updating: Prior to Posterior Analysis

The variance of the sum has the more complicated form

$$V[Z_{t+k}|D_t] = \sum_{j=1}^{k} V[Y_{t+j}|D_t] + 2\sum_{j=1}^{k}\sum_{i=j+1}^{k} \text{Cov}[Y_{t+i}, Y_{t+j}|D_t]$$

$$= \sum_{j=1}^{k} Q_t(j) + 2\sum_{j=1}^{k-1}\sum_{i=j+1}^{k} Q_t(i,j),$$

where, for $i \geq j$, the covariances are defined by

$$Q_t(i,j) = F'_{t+i}C_t(i,j)F_{t+j},$$
$$C_t(i,j) = G^{i-j}R_t(j).$$

To see this latter result consider two future values of the observation series,

$$Y_{t+i} = F'_{t+i}\theta_{t+i} + \nu_{t+i},$$
$$Y_{t+j} = F'_{t+j}\theta_{t+j} + \nu_{t+j},$$

where it is assumed that $i > j$. The covariance between these two quantities is obtained as follows (remembering the independence assumptions on the variance series $\{\nu_t\}$):

$$\begin{aligned}
\text{Cov}[Y_{t+i}, Y_{t+j}] &= \text{Cov}[F'_{t+i}\theta_{t+i} + \nu_{t+i}, F'_{t+j}\theta_{t+j} + \nu_{t+j}] \\
&= \text{Cov}[F'_{t+i}\theta_{t+i}, F'_{t+j}\theta_{t+j}] + \text{Cov}[F'_{t+i}\theta_{t+i}, \nu_{t+j}] \\
&\quad + \text{Cov}[\nu_{t+i}, F'_{t+j}\theta_{t+j}] + \text{Cov}[\nu_{t+i}, \nu_{t+j}] \\
&= F'_{t+i}\text{Cov}[\theta_{t+i}, \theta_{t+j}]F_{t+j} \\
&= F'_{t+i}C_t(i,j)F_{t+j}.
\end{aligned}$$

The covariance between the state at the two times is most easily derived by application of the system equation as in

$$\begin{aligned}
\theta_{t+i} &= G\theta_{t+i-1} + \omega_{t+i} \\
&= G(G\theta_{t+i-2} + \omega_{t+i-1}) + \omega_{t+i} \\
&= \cdots = G^{i-j}\theta_{t+j} + W_t(i,j),
\end{aligned}$$

where $W_t(i,j)$ is a linear combination of the innovation terms for times $t+j+1, ..., t+i$. Now the required state covariance is

$$\begin{aligned}
\text{Cov}[\theta_{t+i}, \theta_{t+j}] &= \text{Cov}[G^{i-j}\theta_{t+j} + W_t(i,j), \theta_{t+j}] \\
&= G^{i-j}V[\theta_{t+j}] \\
&= G^{i-j}R_t(j),
\end{aligned}$$

where the j step ahead state forecast variance $R_t(j)$ has already been defined.

We have assumed in the above derivation that the system matrix is constant. For time varying G the more general recurrence function for step ahead state forecast covariances must be used: $C_t(i,j) = G_{t+i}C_t(i-1,j)$ with $C_t(j,j) = R_t(j)$.

3.2.4 Likelihood

The model likelihood, a function of the model parameters, is the conditional forecast distribution evaluated at the observed value. It has the normal form
$$L(\theta_t | Y_y = y_t, V_t) \propto p(Y_t = y_t | \theta_t, V_t)$$
$$\sim N[F_t'\theta_t, V_t].$$

3.2.5 Posterior Information

The prior information is combined with information in the observation (the likelihood) using Bayes' theorem to yield the posterior distribution on the state,
$$p(\theta_t | D_{t-1}, y_t) = \frac{p(Y_t = y_t | \theta_t, V_t) p(\theta_t | D_{t-1})}{p(Y_t = y_t)}.$$
Notice that the denominator $p(Y_t = y_t)$ is not a function of the state θ_t and serves merely as a density normalising constant. It is therefore typically ignored (since it is easily recovered when the scaled density is required) and Bayesian updating expressed as a proportional form,

$$\text{posterior} \propto \text{likelihood} \times \text{prior}.$$

For the dynamic linear model the state posterior is the product of two normal density functions as just seen, yielding another normal density:
$$p(\theta_t | D_{t-1}, y_t) \propto \exp\{-0.5 V_t^{-1}(y_t - F_t'\theta_t)^2\}$$
$$\times \exp\{-0.5(\theta_t - a_t)' R_t^{-1}(\theta_t - a_t)\}$$
$$\propto \exp\{-0.5(\theta_t - m_t)' C_t^{-1}(\theta_t - m_t)\},$$
where the defining moments are obtained as
$$m_t = a_t + A_t e_t,$$
$$C_t = R_t - A_t A_t' Q_t,$$
$$A_t = R_t F_t / Q_t,$$
$$e_t = y_t - f_t.$$

3.2 Updating: Prior to Posterior Analysis

(Details in West and Harrison, 1989, Chapter 4.) The posterior mean is adjusted from the prior value by a multiple of the one step ahead forecast error. The amount of that adjustment is determined by the quantity A_t which is the regression matrix of the state vector θ_t on the observation Y_t conditional upon the history D_{t-1}. This regression matrix, or adaptive factor as it is called, is determined by the relative size of the state prior variance and the observation variance (we saw above that the forecast variance Q_t is a function of the prior and observation variances). This means that the larger the observation variance compared with the state prior variance, the smaller will be the adaptive factor. The intuition here is that if the observation variance is 'large', then commensurately large forecast errors are quite compatible with the model quantification. Therefore the underlying state estimate is adequate and does not require substantial movement. Conversely, if the state prior variance is large compared with the observation variance, then the observation has a lot of relevant information for the state and adjustment from the prior to the posterior should properly reflect that position.

State posterior variances are smaller than the corresponding prior values because the information base is extended with the new observation. One exception to this rule is the case of a zero regression vector, $F_t = 0$, when the observation Y_t is completely uninformative on the state and the posterior variance is identical to the prior. This applies only to regression components since trends and seasonal components have constant, nonzero regression vectors.

Missing Data

Missing observations are routinely handled in the Bayesian context. Posterior beliefs are simply equal to prior beliefs in the absence of new information. Formally, if y_t is missing, or unreliable and construed to represent no useful information, then $p(\theta_t|D_t) = p(\theta_t|D_{t-1})$.

Aggregate Data

Data arising in the form of aggregates over several periods rather than as individual period values require a minor modification of the prior to posterior analysis. Suppose that total sales for two weeks $(t-1)$ and t is available but that the individual weekly figures are not. Then the state posterior information for time $(t-1)$ is just the prior information for that time: the observation y_{t-1} is missing, so the missing data procedure just described is applied. The posterior for time t is updated with the aggregate information as follows.

Define $Z_t = Y_t + Y_{t-1}$; then from the observation and system equations the sum is expressible in terms of quantities with known distributions,

$$\begin{aligned} Z_t &= F_t'\theta_t + F_{t-1}'\theta_{t-1} + \nu_t + \nu_{t-1} \\ &= F_t'(G\theta_{t-1} + \omega_t) + F_{t-1}'(G\theta_{t-2} + \omega_{t-1}) + \nu_t + \nu_{t-1} \\ &= F_t'[G(G\theta_{t-2} + \omega_{t-1}) + \omega_t] + F_{t-1}'(G\theta_{t-2} + \omega_{t-1}) + \nu_t + \nu_{t-1} \\ &= A_t'\theta_{t-2} + B_t'\omega_{t-1} + F_t'\omega_t + \nu_t + \nu_{t-1}, \end{aligned}$$

with

$$\begin{aligned} A_t' &= (F_{t-1}'G + F_t'G^2), \\ B_t' &= (F_{t-1}' + F_t'G). \end{aligned}$$

Similarly the state vector for time t can be expressed in terms of the posterior for the most recently recorded observation,

$$\theta_t = G^2\theta_{t-2} + G\omega_{t-1} + \omega_t.$$

The joint distribution of the observation sum and the state is normal (since both are marginally normal) and the moments are identifiable from the expressions just derived. This joint distribution is

$$\left. \begin{matrix} Z_t \\ \theta_t \end{matrix} \right| D_{t-2} \sim N\left[\begin{pmatrix} \mu_{t*2} \\ m_{t*2} \end{pmatrix}, \begin{pmatrix} Q_{t*2} & S_{t*2}' \\ S_{t*2} & C_{t*2} \end{pmatrix} \right],$$

where

$$\begin{aligned} \mu_{t*2} &= A_t' m_{t-2}, \\ m_{t*2} &= G^2 m_{t-2}, \\ Q_{t*2} &= A_t' C_{t-2} A_t + B_t' W_{t-1} B_t + F_t' W_t F_t + V_{t-1} + V_t, \\ C_{t*2} &= G^2 C_{t-2} G^{2'} + G W_{t-1} G' + W_t, \\ S_{t*2} &= A_t' C_{t-2} G^{2'} + B_t' W_{t-1} G' + F_t' W_t, \end{aligned}$$

with A_t and B_t defined above. Conditioning θ_t on Z_t now gives the conditional distribution of the state given the aggregate series value and the previous information. Standard normal theory determines this to be

$$(\theta_t | Z_t, D_{t-2}) = (\theta_t | D_t) \sim N[m_t, C_t],$$

where the moments are given by

$$\begin{aligned} m_t &= m_{t*2} + S_{t*2} Q_{t*2}^{-1} [Z_t - \mu_{t*2}], \\ C_t &= C_{t*2} - S_{t*2} Q_{t*2}^{-1} S_{t*2}'. \end{aligned}$$

3.3 Forward Intervention

You may notice that these expressions for the posterior moments have a structure analogous to the usual one step updating formulas. For example, the mean is adjusted from the value implied by its previous posterior (in this case the two periods back posterior) by a proportion of the aggregate forecast error.

Updating with individual observations or observation aggregates is properly consistent. In the two period aggregate illustration just examined, the posterior $p(\theta_t|D_t)$ derived there is exactly the same as the posterior that would ensue if the individual values for Y_{t-1} and Y_t were available and the usual one step Bayes learning procedure applied twice. (Assuming, of course, no intervention at either of these times.)

3.2.6 Evolution

Once an observation is made, and posterior descriptions calculated, concern moves to consideration of the next time. Given the posterior distribution for the state at time $t-1$ as normally distributed with mean m_{t-1} and covariance C_{t-1}, direct application of the system evolution equation leads to the prior for time t. Once again a linear combination of normally distributed quantities yields a normal distribution,

$$\theta_t|D_{t-1} \sim N[a_t, R_t],$$

where the moments are defined by

$$a_t = Gm_{t-1},$$
$$R_t = GC_{t-1}G' + W_t.$$

We have now completed the cycle of prior to forecast to posterior to next prior. These stages characterise the routine on-line updating analysis of the DLM. The analysis is summarised in Table 3.1.

3.3 Forward Intervention

Intervening in an on-line analysis, as described in Section 2.3 of Chapter 2, is achieved by adjusting the model based prior $p(\theta_t|D_{t-1})$ to a new prior $p(\theta_t|D_{t-1}, I_t)$. The example used a doubling of sales level following production difficulties besetting a competitor. If our routine prior belief on sales level is 200,

$$\theta_t|D_{t-1} \sim N[200, R_t],$$

and we double it to 400, then our post-intervention prior is determined directly as

$$\theta_t|D_{t-1}, I_t \sim N[400, R_t],$$

(and typically we would increase the prior variance too, reflecting greater uncertainty). Whatever the values decided upon, we simply have a new quantitative description of the prior on which to base future forecasts.

Formally, subjective intervention of this nature can be modelled by including an additional stochastic term in the system evolution,

$$\theta_t = G\theta_{t-1} + \omega_t + \zeta_t, \qquad \zeta_t \sim N[h_t, H_t],$$

where the intervention variable ζ_t is uncorrelated with the state θ_{t-1} and system innovation ω_t. This extended form of the system equation retains a linear normal structure so the post-intervention prior is normal—only the defining moments are altered.

In our example the state vector includes the level as one element, with price regression as another. Since only the level is affected by the intervention, part of the intervention variable will be identically zero, reflecting no change. Identifying the state as

$$\theta = \begin{pmatrix} \text{level} \\ \text{price coef} \end{pmatrix},$$

the level increase intervention may be formulated by setting the intervention variable moments to be

$$h_t = \begin{pmatrix} 200 \\ 0 \end{pmatrix}, \qquad H_t = \begin{pmatrix} 0 & 0 \\ 0 & 0 \end{pmatrix}.$$

Once again we emphasise that a typical intervention of this kind will result in increased uncertainty *about the component that is the subject of the intervention* so that $H_t(1,1)$ will usually be nonzero. In practice the intervention variable moments are not identified explicitly, rather they are implied by a direct specification of what the post-intervention prior state moments are to be. The general intervention strategy in multicomponent models is discussed in the section on block structured models.

3.3.1 Arbitrary Intervention

The formal intervention mechanism just described—an extended system equation—imposes a restriction on the nature of the allowable intervention. Any desired adjustment of the state mean is possible, but uncertainty can only be increased. For the most part, circumstances necessitating intervention involve increased uncertainty, but not always. It may happen that external information would lead to more certainty on some

3.3 Forward Intervention

aspects of a model. Clearly one may specify a reduced variance for a post-intervention prior from which forecasts and model updating can proceed as usual. A problem arises, however, when calculating retrospective estimates. Smoothing requires a full joint distribution of the state posterior and prior conditional on the intervention information, $p(\theta_{t-1}, \theta_t | D_{t-1}, I_t)$. To carry out the smoothing analysis we need a way of specifying an arbitrary intervention in terms compatible with the standard formulation of the DLM. This is achieved using the following results.

Lemma Let L_t be an n-square lower triangular nonsingular matrix, and h_t any n-vector. Define

$$\theta_t^* = L_t \theta_t + h_t,$$

where $\mathrm{E}[\theta_t] = a_t$ and $\mathrm{V}[\theta_t] = R_t$. For specified moments for θ_t^*, $\mathrm{E}[\theta_t^*] = a_t^*$ and $\mathrm{V}[\theta_t^*] = R_t^*$, set L_t and h_t as follows:

$$L_t = (Z_t^{-1} U_t)',$$
$$h_t = a_t^* - L_t a_t,$$

where U_t and Z_t are the unique upper triangular nonsingular square root matrices (Cholesky factorisation) of R_t^* and R_t respectively, $R_t^* = U_t' U_t$ and $R_t = Z_t' Z_t$.

Proof Omitted.

Note: This is Lemma 11.1 of West and Harrison (1989). Notice, though, the presentation here uses the conventional Cholesky LU decomposition, whereas West and Harrison use a UL decomposition.

Theorem Given the post-intervention prior state mean and variance a_t^* and R_t^*, define L_t and h_t as in the preceding lemma. Then the post-intervention prior is the prior obtained in the DLM with modified system evolution equation

$$\theta_t = G_t^* \theta_{t-1} + \omega_t^*, \qquad \omega_t^* \sim N[h_t, W_t^*],$$

where, given the existing historical information D_{t-1} and the intervention information I_t, the evolution innovation ω_t^* is uncorrelated with the state θ_{t-1} and

$$G_t^* = L_t G,$$
$$\omega_t^* = L_t \omega_t + h_t,$$
$$W_t^* = L_t W_t L_t'.$$

Proof Omitted.

Note: This is Theorem 11.2 of West and Harrison (1989) modified as in the lemma to use the familiar Cholesky LU factorisation.

Smoothing makes use of the modified system matrix G_t^* at intervention times, the basic algorithm and recurrence equations being otherwise unchanged.

3.4 Smoothing

Time series analysis makes use of fitted model estimates using information from all observations to give a view of component development over time. For a data set comprising observations from time 1 to time n this retrospective view utilises the filtered distributions $p(\theta_t|D_n)$ for $t = 1, ..., n$. Information from observations subsequent to a given time is filtered back to that time in a manner similar to the forward filtering of the on-line analysis. The information filtering mechanism is once again Bayes' theorem, with the system equation providing the necessary cross-temporal joint distributions.

To fix ideas consider the case of filtering information from today back to yesterday, from time t to time $t - 1$. The on-line analysis defines prior and posterior distributions for the state at both times,

$$\theta_{t-1}|D_{t-2} \sim N[a_{t-1}, R_{t-1}],$$
$$\theta_{t-1}|D_{t-1} \sim N[m_{t-1}, C_{t-1}],$$
$$\theta_t|D_{t-1} \sim N[a_t, R_t],$$
$$\theta_t|D_t \sim N[m_t, C_t].$$

Attention is now directed toward the retrospective estimate for yesterday, $p(\theta_{t-1}|D_t)$. The derivation proceeds as follows. First, express the filtered distribution in terms of the joint distribution of the state for yesterday and today,

$$p(\theta_{t-1}|D_t) = \int p(\theta_{t-1}|\theta_t, D_t) p(\theta_t|D_t) d\theta_t.$$

The second term in the integrand is today's posterior on the state which we already know. The first term in the integrand is not available directly. It is obtained by applying Bayes' theorem. To begin, separate today's observation y_t from the development of the series up to yesterday D_{t-1}, using $D_t = \{D_{t-1}, y_t\}$,

$$p(\theta_{t-1}|\theta_t, D_t) = p(\theta_{t-1}|\theta_t, y_t, D_{t-1}).$$

Now apply Bayes' theorem to reverse the conditioning on y_t,

$$p(\theta_{t-1}|\theta_t, y_t, D_{t-1}) = \frac{p(y_t|\theta_{t-1}, \theta_t, D_{t-1}) p(\theta_{t-1}|\theta_t, D_{t-1})}{p(y_t|\theta_t, D_{t-1})}.$$

3.4 Smoothing

Given the value of the state today, θ_t, today's observation, y_t, is independent of the value of the state at any previous time, in particular the value yesterday. This is clear from the observation equation. The first term of the numerator is therefore equal to $p(y_t|\theta_t, D_{t-1})$, which cancels with the denominator. The conditional filtered density for the state yesterday is thereby reduced to

$$p(\theta_{t-1}|\theta_t, D_t) = p(\theta_{t-1}|\theta_t, D_{t-1}).$$

In other words, if we know the value of the state today, today's observation provides no additional information on the value of the state yesterday. Heuristically the reasoning is this: Today's observed sales tells us about today's underlying sales level (and any other components in the model) through the observation equation. Today's level then tells us the implications for yesterday's level through the system equation.

We now apply Bayes' theorem once more to reverse the conditioning on θ_{t-1} and θ_t to get the state estimates in the correct temporal ordering,

$$p(\theta_{t-1}|\theta_t, D_{t-1}) = \frac{p(\theta_t|\theta_{t-1}, D_{t-1})p(\theta_{t-1}|D_{t-1})}{p(\theta_t|D_{t-1})}.$$

Now all of the distributional components are known. The first term in the numerator is given directly by the system equation

$$\theta_t|\theta_{t-1}, D_{t-1} \sim N[G\theta_{t-1}, W_t].$$

The second term in the numerator is just the on-line posterior for the state yesterday given at the beginning of the section,

$$\theta_{t-1}|D_{t-1} \sim N[m_{t-1}, C_{t-1}].$$

The denominator is the on-line prior for the state today, also given above,

$$\theta_t|D_{t-1} \sim N[a_t, R_t].$$

It is now a straightforward application of multivariate normal theory to deduce that the conditional filtered distribution is normal,

$$\theta_{t-1}|\theta_t, D_{t-1} \sim N[h_t(1), H_t(1)],$$

where the mean and variance are defined by

$$h_t(1) = m_{t-1} + B_{t-1}(\theta_t - a_t),$$
$$H_t(1) = C_{t-1} - B_{t-1}R_t B'_{t-1},$$
$$B_t = C_t G' R_{t+1}^{-1}.$$

The final calculation required for the one step back filtered estimate is the substitution of the first term of the integrand at the beginning of the section with the distribution just derived,

$$p(\theta_{t-1}|D_t) = \int p(\theta_{t-1}|\theta_t, D_t)p(\theta_t|D_t)\mathrm{d}\theta_t$$
$$= \int p(\theta_{t-1}|\theta_t, D_{t-1})p(\theta_t|D_t)\mathrm{d}\theta_t.$$

Both distributions in the integrand are normal, so standard theory provides the result of the integration directly,

$$\theta_{t-1}|D_t \sim N[a_t(-1), R_t(-1)],$$

where the defining moments are given by

$$a_t(-1) = m_{t-1} + B_{t-1}(m_t - a_t),$$
$$R_t(-1) = C_{t-1} - B_{t-1}(R_t - C_t)B'_{t-1}$$

(and B_{t-1} is defined above).

Filtering back more than one period proceeds in the same way; the general result is given in West and Harrison (1989), Theorem 4.4.

3.5 Component Forms

The formal analysis detailed in this chapter thus far is framed in terms of the general DLM. The updating and smoothing recurrences derived apply quite generally, regardless of the detailed internal structure of a model. Building a suitable model for a time series is facilitated by knowledge of such model structure. Several component forms are discussed and examined from an applied viewpoint in the remainder of the text; here we outline the mathematical structure of those forms.

3.5.1 Polynomial Trend Components

The simplest trend model is the first order polynomial trend, or level. Observed series values are stochastically distributed about a time varying value:

$$Y_t = \mu_t + \nu_t,$$
$$\mu_t = \mu_{t-1} + \omega_t.$$

3.5 Component Forms

Table 3.1 Summary of known variance model

Univariate DLM: known variance V_t		
Observation:	$Y_t = F'_t \theta_t + \nu_t$	$\nu_t \sim N[0, V_t]$
System:	$\theta_t = G_t \theta_{t-1} + \omega_t$	$\omega_t \sim N[0, W_t]$
Information:	\multicolumn{2}{l	}{$(\theta_{t-1} \mid D_{t-1}) \sim N[m_{t-1}, C_{t-1}]$}

$$(\theta_{t-1} \mid D_{t-1}) \sim N[m_{t-1}, C_{t-1}]$$
$$(\theta_t \mid D_{t-1}) \sim N[a_t, R_t]$$
$$a_t = G_t m_{t-1}$$
$$R_t = G_t C_{t-1} G'_t + W_t$$

Forecast:
$$(Y_t \mid D_{t-1}) \sim N[f_t, Q_t]$$
$$f_t = F'_t a_t$$
$$Q_t = F'_t R_t F_t + V_t$$

Updating Recurrence Relationships

$$(\theta_t \mid D_t) \sim N[m_t, C_t]$$
$$m_t = a_t + A_t e_t$$
$$C_t = R_t - A_t A'_t Q_t$$
$$e_t = Y_t - f_t$$
$$A_t = R_t F_t / Q_t$$

Forecast Distributions

For $k \geq 1$,
$$(\theta_{t+k} \mid D_t) \sim N[a_t(k), R_t(k)]$$
$$(Y_{t+k} \mid D_t) \sim N[f_t(k), Q_t(k)]$$
$$a_t(k) = G_{t+k} a_t(k-1)$$
$$R_t(k) = G_{t+k} R_t(k-1) G'_{t+k} + W_{t+k}$$
$$f_t(k) = F'_{t+k} a_t(k)$$
$$Q_t(k) = F'_{t+k} R_t(k) F_{t+k} + V_{t+k}$$
$$a_t(0) = m_t$$
$$R_t(0) = C_t$$

Filtered Distributions

For $1 \leq k \leq t$,
$$(\theta_{t-k} \mid D_t) \sim N[a_t(-k), R_t(-k)]$$
$$a_t(-k) = m_{t-k} - B_{t-k}[a_{t-k+1} - a_t(-k+1)]$$
$$R_t(-k) = C_{t-k} - B_{t-k}[R_{t-k+1} - R_t(-k+1)]B'_{t-k}$$
$$B_t = C_t G'_{t+1} R^{-1}_{t+1}$$

Intervention Adjustment

Pre-intervention prior: $(\theta_t \mid D_{t-1}) \sim N[a_t, R_t]$
Post-intervention prior: $(\theta_t \mid D_{t-1}, I_t) \sim N[a^*_t, R^*_t]$
Define U_t, Z_t by $R^*_t = U'_t U_t$, $R_t = Z'_t Z_t$
Set $L_t = (Z_t^{-1} U_t)'$, $h_t = a^*_t - L_t a_t$, $G^*_t = L_t G_t$
Then G^*_t, R^*_t, a^*_t replace G_t, R_t, a_t in filtered equations.

The system equation defines the level to be a simple random walk through time. Identifying the terms of a defining DLM it is easy to see that the regression vector and system evolution matrix are constants,

$$F_t = 1, \quad G = 1.$$

A second order polynomial trend model allows for systematic growth or decline in level. The additional parameter quantifies the time to time change in level,

$$Y_t = \mu_t + \nu_t,$$
$$\mu_t = \mu_{t-1} + \beta_{t-1} + \omega_{1t},$$
$$\beta_t = \beta_{t-1} + \omega_{2t}.$$

The state vector comprises two elements, $\theta_t = (\mu_t, \beta_t)'$, the first representing the current level and the second representing the current rate of change in the level. Identifying the regression vector and system matrix is again straightforward,

$$F_t = \begin{pmatrix} 1 \\ 0 \end{pmatrix}, \quad G = \begin{pmatrix} 1 & 1 \\ 0 & 1 \end{pmatrix}.$$

Higher order polynomial time trend models may be defined by generalising and extending the first and second order models described here. Practically speaking, such trend descriptions are seldom used and are not discussed further in this book. West and Harrison (1989), Chapter 7, provides details; Pole and West (1990) includes an example of a quadratic growth (third order polynomial) model.

3.5.2 Seasonal Component Models

Modelling seasonal patterns in time series requires a component form that is periodic. The most natural representation of such patterns is the seasonal factor model, where a different parameter—a factor—is defined for each point in a cycle. For example, quarterly data exhibiting an annual cycle is modelled with a set of four factors, one for each quarter.

An alternative representation isolates an underlying trend from periodic movement about that trend. Over a complete cycle the effects sum to zero since the trend, which is just the average of the factors over the cycle, contains the overall series movement over that time span. This seasonal effects model defines parameters to measure seasonal departures from a trend. To illustrate, a set of seasonal factors 100, 140, 80, and 120 is equivalent to a trend of 110 and seasonal effects -10, 30, -30, and 10. In practice, the effects model is preferred over the factors model because it facilitates separation of underlying trend and seasonal variation about

3.5 Component Forms

the trend. The trend, of course, may have any of the forms discussed previously. For example, cyclical behaviour may be superimposed on a generally increasing level (see Figure 2.1 in Chapter 2).

For the quarterly, annual cycle, seasonal effect DLM the system state comprises four effects parameters, one for each quarter. As time proceeds, the state vector simply rotates by one element each period. If the last quarter was quarter three, then the state is ordered as

$$\theta_{t-1} = \begin{pmatrix} \text{qtr 3} \\ \text{qtr 4} \\ \text{qtr 1} \\ \text{qtr 2} \end{pmatrix}.$$

This quarter (fourth) the state is rotated once to give the new ordering

$$\theta_t = \begin{pmatrix} \text{qtr 4} \\ \text{qtr 1} \\ \text{qtr 2} \\ \text{qtr 3} \end{pmatrix}.$$

The cycling state is expressible in the DLM system equation by setting the evolution matrix to have the cyclic form

$$G = \begin{pmatrix} 0 & 1 & 0 & 0 \\ 0 & 0 & 1 & 0 \\ 0 & 0 & 0 & 1 \\ 1 & 0 & 0 & 0 \end{pmatrix}.$$

To complete the model definition the observation equation simply picks off the current quarter effect from the state so the regression vector is simply $F_t = \begin{pmatrix} 1 & 0 & 0 & 0 \end{pmatrix}'$.

For a general seasonal component with period p the definitions of the regression vector and system matrix extend naturally to

$$F_t = \begin{pmatrix} 1 \\ 0 \\ \vdots \\ 0 \\ 0 \end{pmatrix}, \quad G = \begin{pmatrix} 0 & 1 & 0 & 0 & \cdots & 0 \\ 0 & 0 & 1 & 0 & \cdots & 0 \\ \vdots & & & & & \vdots \\ 0 & 0 & 0 & 0 & \cdots & 1 \\ 1 & 0 & 0 & 0 & \cdots & 0 \end{pmatrix}.$$

With a DLM seasonal component model structured in this effect form, there is one complicating factor in the statistical analysis. The zero sum constraint on the effects must be maintained in all of the probabilistic descriptions, priors and posteriors, otherwise the model is difficult to interpret. Without the constraint the underlying trend becomes confounded with the average of the 'effects'—which are no longer effects, of course.

If the constraint is satisfied by the prior distribution, then it is guaranteed to be satisfied by the posterior distribution by the rules of the probability calculus. A constraint is just a condition. Unless it is explicitly removed, it is maintained through Bayes' theorem. Let C denote the constraint; then the posterior is given as

$$p(\theta_t|C, D_{t-1}, y_t) \propto p(y_t|\theta_t, D_{t-1}, C)p(\theta_t|C, D_{t-1})$$
$$\propto p(y_t|\theta_t)p(\theta_t|C, D_{t-1}).$$

(Given the state at time t the observation is independent of previous series history and the constraint; the constraint, of course, is already implicit in the value of the state on which the conditioning is defined.) Formal proof may also be seen through the updating equations. The state posterior mean and variance can be expressed as

$$m_t = a_t + R_t F_t Q_t^{-1} e_t,$$
$$C_t = R_t - R_t F_t Q_t^{-1} F_t' R_t'$$

(substituting for the adaptive factor in the equations in Table 3.1). The zero sum constraint $1'\theta = 0$ is satisfied by the prior distribution if the mean and variance satisfy $1'a_t = 0$ and $R_t 1 = 0$ respectively. Without complication it can be seen from the equations above that if the prior moments satisfy the aforementioned conditions, then the posterior moments also satisfy those conditions, $1'm_t = 0$ and $C_t 1 = 0$, and hence the posterior distribution satisfies the constraint.

It remains only to ensure that the prior description of a seasonal component satisfies the constraint at any time. Priors are either directly specified by the investigator through intervention or are derived from the previous period's posterior by way of the state evolution. In direct specification, individual seasonal effect priors are typically quantified independently. Moreover, it is usual for information to be limited to a description of the marginals—moments of individual effects—with no information forthcoming on covariances between pairs of effects. In that circumstance the zero sum constraint must be applied in constructing the joint prior distribution for the set of effects. The following result provides the means by which the constraint is imposed.

3.5 Component Forms

Theorem *Given an initial prior* $\theta \sim N[a^*, R^*]$ *with arbitrary moments* a^* *and* R^*, *imposing the constraint* $1'\theta = 0$ *leads to the revised prior* $\theta \sim N[a, R]$ *where*
$$a = a^* - A(1'a^*)/(1'R^*1),$$
$$R = R^* - AA'/(1'R^*1),$$
and $A = R^*1$.

Proof Omitted. See West and Harrison (1989), Theorem 8.2.

The prior distribution obtained from the system evolution applied to the previous posterior will satisfy the constraint if the innovation term ω_t meets the constraint conditions. (The system matrix G is just a permutation matrix which serves only to reorder the individual state elements, so it does not affect the relationship between effects.) This term has zero mean, so attention need only focus on the variance W_t. In practice this variance is specified using the discount approach so that W_t is a fraction of the previous posterior variance C_{t-1}. Since the posterior distribution satisfies the condition $C_{t-1}1 = 0$, so does the innovation, $W_t 1 = (\delta^{-1} - 1)C_{t-1}1 = 0$.

3.5.3 Harmonic Analysis

Mathematical functions such as sines and cosines exhibit well known cyclical behaviour. The cosine function $\cos(\omega(t-1))$ defines a periodic function with period $2\pi/\omega$. When $\omega = \pi/6$ the cosine function has period 12 as shown in the lower panel of Figure 3.1. If a monthly time series with an annual seasonal pattern is well approximated by the cosine form, then a very parsimonious model can be used in modelling. Instead of 11 effect parameters (one for each month plus the zero sum constraint) only one is required,
$$Y_t = a_t \cos\left(\frac{\pi(t-1)}{6}\right) + \nu_t.$$

The parameter a_t models an arbitrary amplitude (seasonal peak to trough variation); in Figure 3.1, a_t is unity. A simple generalisation of the basic cosine wave allows the cycle maximum and minimum to be phase shifted. The seasonal peak can be translated from the default January position to any other month by using a sum of sine and cosine terms with the same frequency,
$$Y_t = a_t \cos\left(\frac{\pi(t-1)}{6}\right) + b_t \sin\left(\frac{\pi(t-1)}{6}\right) + \nu_t.$$

This model has two parameters, combinations of which may be defined to give a cosine wave of period 12 with any desired amplitude and phase shift.

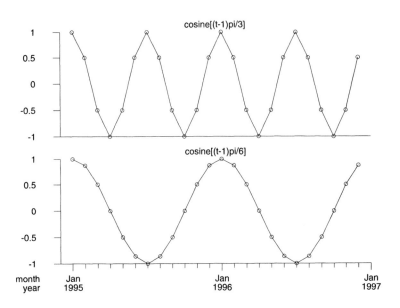

Figure 3.1 Cosine functions with periods 6 and 12 months.

A DLM formulation of this single harmonic function is obtained using a two parameter state, with the regression vector and system matrix defined as follows:

$$F_t = \begin{pmatrix} 1 \\ 0 \end{pmatrix}, \qquad G = \begin{pmatrix} \cos\omega & \sin\omega \\ -\sin\omega & \cos\omega \end{pmatrix}.$$

The system matrix is p-cyclic: like the permutation matrix used in the seasonal effects model it repeats itself after p periods, $G^p = G$.

More complicated cyclical patterns are obtained by considering additional cosine waves with higher frequencies. The function $\cos(2\omega(t-1))$ is similar to the function just examined with the exception that it completes two full cycles in the time it takes the first to complete one. Its period is half that of the first or, equivalently, its frequency is twice as great. With $\omega = \pi/6$ this higher frequency function completes two full cycles in 12 time periods, Figure 3.1 upper panel. Once again a linear combination of sine and cosine terms at this frequency allows for an arbitrary peak to trough range and phase shift.

Adding two such periodic functions together, Figure 3.2, enables a much wider range of cyclical behaviour to be modelled than is possible with either

3.5 Component Forms

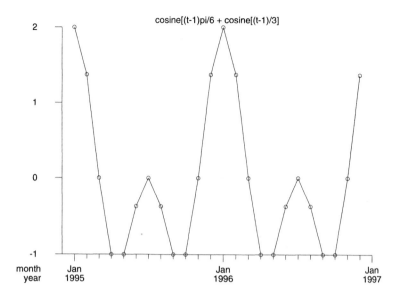

Figure 3.2 Sum of two cosine functions with periods 6 and 12 months.

function individually—and still with only four parameters,

$$Y_t = [a_1 t \cos \omega(t-1) + b_1 t \sin \omega(t-1)]$$
$$+ [a_2 t \cos 2\omega(t-1) + b_2 t \sin 2\omega(t-1)] + \nu_t.$$

The regression vector and system matrix for this two harmonic model are

$$F_t = \begin{pmatrix} 1 \\ 0 \\ 1 \\ 0 \end{pmatrix}, \quad G = \begin{pmatrix} \cos \omega & \sin \omega & 0 & 0 \\ -\sin \omega & \cos \omega & 0 & 0 \\ 0 & 0 & \cos 2\omega & \sin 2\omega \\ 0 & 0 & -\sin 2\omega & \cos 2\omega \end{pmatrix}.$$

Continuing to add higher frequency terms more and more complex seasonal patterns come within the ambit of the harmonic formulation. With a full set of harmonics any arbitrary pattern is representable as the Fourier Representation theorem shows.

Theorem *Fourier Representation theorem: Any cyclical function of period p defined by a set of p effects $\phi_1, ..., \phi_p$ can be expressed as a linear combination of sine and cosine terms. Define $\omega = 2\pi/p$; then there exist $(p-1)$ real numbers $a_1, ..., a_q; b_1, ..., b_{q-1}$ such that, for $j = 1, ..., p$,*

$$\phi_j = \sum_{r=1}^{q-1} [a_r \cos(\omega r j) + b_r \sin(\omega r j)] + a_q \cos(\pi j),$$

where $p = 2q$ if p is even, and $p = 2q - 1$ with $a_q = 0$ if p is odd.

Proof *Omitted.* For details, and definitions of the Fourier coefficients a_i, b_j, see West and Harrison (1989), Chapter 8. (Note: Their version of the theorem includes an extra parameter, a_0, as they use the factor model, which includes the cycle level.)

Why Harmonic Analysis?

We noted at the beginning of this subsection that, when appropriate, a restricted harmonic model for seasonal variation can be very parsimonious. In practice, relating the coefficients of sine and cosine terms to the observation series is not easy. In fact, knowledge of the full harmonic machinery is necessary and even then interpretation of the coefficients is fraught with difficulty. Imagine trying to explain to a company sales manager that his product exhibited yearly seasonal variation described by a cosine wave with amplitude such-and-such and phase shift so-and-so.

It should be emphasised that practical application of harmonic models is not carried out in terms of the Fourier coefficients. As technical analysts we may utilise a harmonic model for our convenience, but communication and interpretation is always made in terms of meaningful quantities, namely seasonal effects. See the applications in Chapters 4 and 6, and the tutorial demonstration of seasonal component models in Chapter 10.

Technical details of the relationship between Fourier coefficients and seasonal effects are straightforward. The Fourier representation theorem defines a unique linear transformation from a set of seasonal effects to the coefficients of the set of cycle harmonics. The transformation is invertible so that seasonal effects may be uniquely determined from a Fourier description of a cycle. Define a vector of Fourier coefficients as $\theta_t = (a_1, ..., a_q, b_1, ..., b_{q-1})'$—with a_q identically zero and omitted when p is odd—then the constrained seasonal effects vector ϕ_t is obtained from the linear transformation,

$$\phi_t = L\theta_t.$$

The transformation matrix has dimension $p \times (p-1)$ and is defined as

$$L = \begin{pmatrix} F' \\ F'G \\ \vdots \\ F'G^{p-1} \end{pmatrix},$$

where the regression vector and system matrix correspond to the DLM representation of the harmonic model (above). The reverse transformation—from effects to Fourier coefficients—is given by

$$\theta_t = H\phi_t,$$

3.6 Superposition: Block Structured Models

where H is the $(p-1) \times p$ matrix

$$H = (L'L)^{-1}L'.$$

Details in West and Harrison (1989), Chapter 8.

We have already stressed the economy of representation that derives from a harmonic analysis when less than the full set of harmonics is sufficient to model seasonal patterns. But a warning should be sounded here. If a reduced set harmonic model is used and system dynamics subsequently change, so that an alternative cyclical form is exhibited, then model forecast performance may deteriorate badly. Monitoring forecast performance and careful diagnosis of forecast performance breakdown will enable identification of such problems. Subsequent experimentation and investigation will establish new model forms.

3.5.4 Regression Components

Regressions on exogenous series fit into the DLM in much the same way as in ordinary linear models. A regression on price, for example, looks like

$$Y_t = \alpha_t \text{price}_t + \nu_t,$$
$$\alpha_t = \alpha_{t-1} + \omega_t.$$

The regression coefficient has a simple random walk evolution, exactly the same as the first order polynomial trend discussed above. Regressing on several variables, $X_1, ..., X_q$, has the expected form

$$Y_t = \beta_{1t} X_{1t} + \cdots + \beta_{qt} X_{qt} + \nu_t,$$
$$\beta_{it} = \beta_{it-1} + \omega_{it}, \qquad i = 1, ..., q,$$

being a collection of individual regression components (see the discussion of block structuring below). The DLM regression vector for a regression component is self-evidently the regression variable itself, $F_t = X_t$, and the system matrix is the identity, $G = 1$.

3.6 Superposition: Block Structured Models

Component forms for trend, seasonal, and regression are the building blocks for constructing models of complex time series behaviour. A seasonal effects DLM will typically require a trend component—only if the series has a fixed zero underlying level will this not be so. The linear additive structure of the DLM enables component models to be brought together in a straightforward manner. A linear growth trend plus single harmonic seasonal model is a DLM with regression vector and system matrix defined by

$$\theta_t = \begin{pmatrix} \theta_T \\ \theta_S \end{pmatrix}_t,$$

$$F_t = \begin{pmatrix} F_T \\ F_S \end{pmatrix}_t = \begin{pmatrix} 1 \\ 0 \\ 1 \\ 0 \end{pmatrix},$$

$$G = \begin{pmatrix} G_T & 0 \\ 0 & G_S \end{pmatrix} = \begin{pmatrix} 1 & 1 & 0 & 0 \\ 0 & 1 & 0 & 0 \\ 0 & 0 & \cos\omega & \sin\omega \\ 0 & 0 & -\sin\omega & \cos\omega \end{pmatrix}.$$

Note that the regression vector simply stacks the individual regression vectors on top of each other; similarly the system state stacks up the component states. The system matrix is structured block diagonally. (Naturally, for all three quantities the stacking order must be the same, although the particular order chosen does not matter.)

In this way models with any number of components may be constructed. A more complex seasonal pattern—with multiple harmonics—just adds in more of the harmonic components (with different frequencies); regressions just tag on one at a time.

3.6.1 Block Discounting

Discounting was introduced in Chapter 2 as a practicable solution to the problem of setting evolution disturbance variances. To recap, the state prior variance at any time is computed as a function of the posterior variance at the previous time determined by a discount factor, $\delta \in (0, 1]$. The discount factor represents the amount of information loss attributed to temporal advancement,

$$V[\theta_t|D_{t-1}] = \delta^{-1} G V[\theta_{t-1}|D_{t-1}] G'.$$

Discounting variances this way is equivalent to setting the evolution variance as a proportion of the posterior variance,

$$W_t = \left(\frac{1}{\delta} - 1\right) G C_{t-1} G',$$

recalling that C_{t-1} denotes the posterior variance $V[\theta_{t-1}|D_{t-1}]$.

3.6 Superposition: Block Structured Models

Notice that this representation is more general than the simple one parameter (level) example illustrated in Chapter 2. There, with an identity evolution matrix G the discount factor was applied directly to the previous posterior variance. In the general case discussed here the form of G must be explicitly included. Of course, when $G = I$ (random walk evolutions) the general case simplifies to the special case and G may then be omitted.

In models with multiple components the recommended discount strategy is to proceed component by component. Separate discount factors are specified for each component and individual component evolution variance matrices computed. The overall state evolution variance matrix is then set to the block diagonal composition of these individual elements.

To illustrate, the trend plus seasonal example above has two components. Defining δ_T and δ_S to be the discount factors for these components, structure the evolution covariance matrix as

$$W_t = \begin{pmatrix} W_T & 0 \\ 0 & W_S \end{pmatrix}_t$$

$$= \begin{pmatrix} \left(\frac{1}{\delta_T} - 1\right) G_T C_{Tt-1} G_T' & 0 \\ 0 & \left(\frac{1}{\delta_S} - 1\right) G_S C_{St-1} G_S' \end{pmatrix},$$

where C_{Tt-1} is the posterior covariance (sub)matrix for the trend component at time $(t-1)$, and C_{St-1} is similarly the posterior covariance for the seasonal component.

Structuring the evolution covariance using block discounting has several practical advantages. Crucially important in practice, separate discount factors allow distinct components to evolve more or less quickly—to exhibit, and model, more or less stability. Seasonal variation, for example, is often observed to be more stable over longer periods of time than is trend variation. Setting cross-component correlations in the evolution covariance matrix to be zero means that these same correlations are reduced in the prior (because component variances are increased). Intracomponent parameter correlations are not changed. This procedure reduces the strength of coupling between components allowing movement in such relationships if necessary.

A seasonal component model typically comprises several harmonics. Although the individual harmonics are mathematically fully fledged component forms, they are treated collectively as a description of a single seasonal component. A single seasonal discount factor is used to determine the seasonal evolution variance and it is applied to the covariances of the overall seasonal parameter block, not on a harmonic-by-harmonic basis.

Regression variables may be treated individually for discounting or grouped together. Models with several regressors could conceivably include a number of sets of regressors, each set having its own discount factor and treated as a single component model. Whatever grouping emerges in an application, the evolution covariance matrix is built in the block diagonal form exemplified above.

3.6.2 Component Intervention

The multicomponent structure of the DLM facilitates a decomposition of time series into recognisable constituent parts. It also thereby enables a natural modelling of changes in individual series components. A jump in level is directly modelled by appropriate intervention on the level component. Other series characteristics, represented by other model components, are not affected by such changes.

The recommended intervention procedure calls for the setting to zero of all the correlations between an altered component and all other components. In the trend plus seasonal example, an intervention to double the expected value of the level would leave the seasonal component quantification unchanged, but set the post-intervention prior correlation between trend parameters (level and growth) and seasonal parameters (Fourier coefficients) to zero,

$$V[\theta_t|D_{t-1}, I_t] = \begin{pmatrix} R_{Tt}^\dagger & 0 \\ 0 & R_{St} \end{pmatrix},$$

where $R_{St} = \delta_S^{-1} G_S C_{St-1} G_S'$ and the '$\dagger$' notation denotes a directly set post-intervention value.

The rationale for setting prior correlations to zero in this way is based on the principle of insurance. When we intervene on a component we typically do so because of knowledge of or belief in an impending change. The nature of such knowledge is that, for the most part, uncertainty surrounding the change is great—and we could always be wrong in our expectations. Increased prior variances on a changed component allow that component to adjust more rapidly to incoming data, leaving behind the baggage of now less relevant history. But any such changes should be isolated from other components during this intensive learning phase. Their historical development is still quite valid and we do not want any distortions to arise from mistaken beliefs about the changed component. Isolation is assured by zero prior correlation.

3.7 Variance Learning

The analysis detailed thus far has assumed a known observation variance sequence $\{V_t\}$. In most applications the observation variance is not known sufficiently well to approximate it by a fixed value. A learning mechanism is necessary. A tractable analytical solution exists for this problem, an extension of the normal-gamma conjugate analysis for the standard linear model.

Working in terms of the precision, $\phi = V^{-1}$, define the constant unknown variance DLM as

$$\text{Observation Equation}: \quad Y_t = F'_t \theta_t + \nu_t, \quad \nu_t \sim N[0, \phi^{-1}],$$
$$\text{System Equation}: \quad \theta_t = G_t \theta_{t-1} + \omega_t, \quad \omega_t \sim N[0, W^*_t \phi^{-1}].$$

The scaling of the system disturbance covariance by the unknown observation variance is necessary for a conjugate analysis. It is just a scale factor, of course, and clearly we can set $W_t = W^*_t \phi^{-1}$ to recover the normal form of the system equation. In practice the discount formulation is used to determine W_t so the mathematical niceties here need not be cause for concern.

Prior Information

At time t the prior information extends over both the state and the scale parameter,

$$\theta_t | D_{t-1}, \phi \sim N[a_t, R^*_t \phi^{-1}],$$
$$\phi | D_{t-1} \sim G[n_{t-1}/2, d_{t-1}/2].$$

Conditional on the scale, the prior information and the model specialise to the normal DLM already detailed (taking $R_t = R^*_t \phi^{-1}$). The parameters of the gamma prior on the scale represent the degrees of freedom, n_{t-1}, and sums of squared errors, d_{t-1}, with mean equal to the ratio of these quantities, n_{t-1}/d_{t-1}. The significance of these definitions will become clear later in the analysis.

Forecasts

The conditional forecast distribution is the normal form seen before, now with the scale made explicit,

$$Y_t | D_{t-1}, \phi \sim N[F'_t a_t, \phi^{-1} Q^*_t],$$

where the observation scale free forecast variance is

$$Q^*_t = 1 + F'_t R^*_t F_t.$$

From standard normal-gamma theory, unconditionally the forecast has a t distribution on n_{t-1} degrees of freedom,

$$Y_t|D_{t-1} \sim t_{n_t-1}[f_t, Q_t],$$

where the mean and scale parameter are given in familiar forms,

$$f_t = F_t' a_t,$$
$$Q_t = S_{t-1} + F_t' R_t^* S_{t-1} F_t.$$

Notice the similarity of this result with the known variance case. The point forecast is the same in both cases. The forecast scale has the same algebraic form with the exception that the unknown variance is estimated by its prior expected value, $S_{t-1} = d_{t-1}/n_{t-1}$.

Posterior Information

Obtaining the posterior distribution on the scale parameter is an easy application of Bayes' theorem since we can eliminate the state parameter by using the conditional predictive distribution just described and the marginal prior on the scale:

$$p(\phi|D_{t-1}, y_t) \propto p(y_t|D_{t-1}, \phi) p(\phi|D_{t-1})$$

$$\propto (\phi^{-1} Q^*)^{-1/2} \exp\left\{-\frac{1}{2}(\phi^{-1} Q^*)^{-1/2}(y_t - f_t)^2\right\}$$

$$\times \phi^{n_{t-1}/2 - 1} \exp\left\{-\frac{\phi}{2} d_{t-1}\right\}$$

$$\propto \phi^{(n_{t-1}-1)/2} \exp\left\{-\frac{\phi}{2}\left[\frac{e_t^2}{Q_t^*} + d_{t-1}\right]\right\}$$

$$\sim G[n_t/2, d_t/2],$$

where
$$n_t = n_{t-1} + 1,$$
$$d_t = d_{t-1} + e_t^2/Q_t^*.$$

The scale posterior has the same distributional form as the prior, a gamma distribution, but with updated parameters. The degrees of freedom parameter increases by one—an additional piece of information has been processed—and the sums of squares term is incremented by the square of the (scaled) forecast error. The posterior point estimate of the scale is just the ratio of these quantities, the sums of squares divided by the degrees of freedom, $S_t = d_t/n_t$.

3.7 Variance Learning

The posterior for the state is most easily obtained by considering the joint distribution with the observation. Both conditional distributions, $(Y_t|D_{t-1}, \phi)$ and $(\theta_t|D_{t-1}, \phi)$, are normal, so their joint distribution is also normal, the covariance being easily identified from the observation equation

$$\begin{pmatrix} Y_t \\ \theta_t \end{pmatrix} \Big| D_{t-1}, \phi \sim N \left[\begin{pmatrix} f_t \\ a_t \end{pmatrix}, \begin{pmatrix} \phi^{-1}Q_t^* & F_t'R_t^*\phi^{-1} \\ R_t^*F_t\phi^{-1} & R_t^*\phi^{-1} \end{pmatrix} \right].$$

Using properties of the multivariate normal distribution, conditioning the state on the observed value yields a normal distribution which is directly identified as

$$(\theta_t|D_{t-1}, Y_t = y_t, \phi) \sim N[m_t, C_t^*\phi^{-1}],$$

where the moments are updated from the prior values as already seen for the normal analysis, the difference here being that the scale conditioning is made explicit,

$$m_t = a_t + R_t^* F_t \phi^{-1} (\phi^{-1}Q_t^*)^{-1}(y_t - f_t)$$
$$= a_t + R_t^* F_t e_t / Q_t^*,$$

$$C_t^* \phi^{-1} = R_t^* \phi^{-1} - R_t^* F_t \phi^{-1} (\phi^{-1}Q_t^*)^{-1} F_t' R_t^* \phi^{-1}$$
$$= R_t^* \phi^{-1} - R_t^* F_t F_t' R_t^* \phi^{-1} / Q_t^*.$$

The marginal posterior state distribution—removing the conditioning on the scale parameter—is obtained by integrating over ϕ in the joint distribution,

$$p(\theta_t|D_t) = \int p(\theta_t, \phi|D_t) d\phi$$
$$= \int p(\theta_t|D_t, \phi) p(\phi|D_t) d\phi$$
$$\propto \int |C_t^* \phi^{-1}|^{-\frac{1}{2}}$$
$$\times \exp\left\{-\frac{1}{2}(\theta_t - m_t)'[C_t^*\phi^{-1}]^{-1}(\theta_t - m_t)\right\}$$
$$\times \phi^{n_t/2 - 1} \exp\{-\phi d_t/2\} d\phi$$
$$\int \phi^{(n_t+p)/2 - 1} \exp\left\{-\frac{\phi}{2}\left[(\theta_t - m_t)'C_t^{*-1}(\theta_t - m_t) + d_t\right]\right\} d\phi$$

(where p is the dimension of the state vector, θ.) Writing $S_t = d_t/n_t$ (the point estimate of the scale parameter) the integral reduces to

$$\left[n_t + (\theta_t - m_t)'(C_t^* S_t)^{-1}(\theta_t - m_t)\right]^{-(n_t+p)/2}.$$

This is the form of a multivariate Student-t distribution (in p dimensions) on n_t degrees of freedom with mean m_t and scale $C_t = C_t^* S_t$.

3.7.1 Variance Laws

In certain time series it is observed that random variation about the mean value function exhibits a well defined pattern. A proportionality relationship has been found to be useful in application areas as diverse as product demand (Stevens, 1974) and medical monitoring (Smith and West, 1983). Let $\mu_t = F_t'\theta_t$ denote the series level (be careful not to confuse this with the trend component), then the general 'variance proportional to level' model is expressed with the observation equation

$$Y_t = \mu_t + \nu_t, \qquad \nu_t \sim N[0, \mu_t^i \phi^{-1}],$$

where $i > 0$ is a power law index. The variance of the forecast for Y_t now requires knowledge of the series level itself—but is precisely what is being forecast! A reasonable practical solution to this dilemma is to use the expected value, $E[\mu_t|D_{t-1}]$. This works well in practice because it is really only necessary to know the order of magnitude of the level factor. Smaller discrepancies in a variance law have little practical consequence on forecasting given all the other model uncertainties, particularly uncertainty in the scale parameter, ϕ. The approximate model therefore has the usual conditional forecast distribution

$$Y_t|D_{t-1}, \phi \sim N[f_t, Q_t],$$

but now with variance

$$Q_t = F_t' R_t F_t + f_t^i \phi.$$

The unconditional distribution has a similar alteration to its scale parameter,

$$Y_t|D_{t-1} \sim t_{n_{t-1}}[f_t, Q_t],$$

with

$$Q_t = F_t' R_t F_t + f_t^i S_{t-1}.$$

Variance laws of many kinds can be modelled with this approach; all that is necessary is a known expression for the scale multiplier (μ_t in the example here). West and Harrison (1989), Chapter 10, gives a more extensive discussion of variance law modelling.

3.7.2 Variance Discounting

The DLM stresses dynamic modelling through time, the stochastic element of the system equation providing for parametric change. Supporting arguments in favour of dynamic change apply equally strongly to the scale

3.7 Variance Learning

parameter as to the system state. The variance learning model analysed above does not allow for such a dynamic—the scale is explicitly determined as an unknown constant. That analysis may be extended to permit stochastic change in the scale using the information discounting strategy.

Posterior information on the scale at time $(t-1)$ is described by a gamma distribution,

$$\phi_{t-1}|D_{t-1} \sim G[n_{t-1}/2, d_{t-1}/2].$$

(Note that we now subscript the scale parameter to distinguish prior from posterior estimates at any time.) For the variance learning analysis to remain tractable the prior information on the scale for time t must stay within the gamma family of distributions. But while the distributional form cannot be changed, the parameterisation may be altered as desired.

The system equation specifies a formal model for stochastic evolution of the state, but that is extremely difficult to do for the scale. The reason lies in the functional form of the gamma distribution—it does not have the convenient mathematical properties the normal distribution exhibits. However, a formal model of scale evolution is not strictly necessary. What matters is to have the time t prior information suitably adjusted from the time $(t-1)$ posterior information so as to encompass temporal uncertainty. Discounting the precision is a natural way of achieving this aim as seen with the state. Applying the idea to the scale parameter leads to discounting both parameters of the gamma distribution. For a variance component discount factor δ define the prior information on the scale as

$$\phi_t|D_{t-1} \sim G[\delta n_{t-1}/2, \delta d_{t-1}/2].$$

The mean of the prior is preserved at the posterior value since it is the ratio of the two parameters, $(\delta n_{t-1}/2)/(\delta d_{t-1}/2) = (n_{t-1}/2)/(d_{t-1}/2)$, while the variance is inflated by a factor of δ^{-1}, just as was done for the state.

Unconditional forecast, state prior and posterior distributions are simple modifications of the undiscounted case (assuming no variance law for simplicity),

$$Y_t|D_{t-1} \sim t_{\overline{n}_t}[f_t, Q_t],$$
$$\theta_t|D_{t-1} \sim t_{\overline{n}_t}[a_t, R_t],$$
$$\theta_t|D_t \sim t_{n_t}[m_t, C_t],$$

where
$$f_t = F'_t a_t,$$
$$Q_t = \overline{S}_t + F'_t R_t F_t,$$
$$m_t = a_t + R_t F_t e_t / Q_t,$$
$$C_t = (S_t/\overline{S}_t)[R_t - R_t F_t F'_t R_t]/Q_t$$
$$= (S_t/\overline{S}_t) R_t [I - F_t A'_t]$$
$$= (S_t/\overline{S}_t)[R_t - A_t A'_t Q_t],$$
$$\overline{n}_t = \delta n_{t-1},$$
$$\overline{d}_t = \delta d_{t-1},$$
$$\overline{S}_t = \overline{n}_t/\overline{d}_t = S_{t-1}.$$

There are four basic points of departure of this unknown variance analysis from the known variance analysis. In the forecast variance Q_t, the prior point estimate $\overline{S}_t$ replaces the previously known value V_t. The state posterior covariance is rescaled at each time to reflect the latest information on the scale parameter (the factor $(S_t/\overline{S}_t)$ in the equation for C_t). The marginal distributions are now Student-t instead of normal. And of course there is the probabilistic description of the scale itself. The scale posterior is clearly the same form as in the no-discount model but with different parameter values,
$$\phi_t|D_t \sim G[n_t/2, d_t/2],$$
where
$$n_t = \overline{n}_t + 1,$$
$$d_t = \overline{d}_t + S_{t-1} e_t^2 / Q_t.$$

3.7.3 Variance Intervention

Subjective intervention on the variance component is performed in the same way as it is on the state components. The routine model based prior, $p(\phi_t|D_{t-1})$, is directly altered to reflect external information, I_t, giving the post-intervention prior, $p(\phi_t|D_{t-1}, I_t)$.

3.7.4 Smoothing

The smoothing analysis for the variance known model extends to the variance learning model in the same way that the on-line analysis extends. Conditional on the scale, all relevant distributions are normal; then the conditioning is removed by integrating over the posterior distribution on

3.7 Variance Learning

ϕ. From Table 3.1 the filtered state distribution is

$$\theta_{t-k}|D_t, \phi \sim N[a_t(-k), R_t(-k)],$$

where, making the conditioning on ϕ explicit,

$$a_t(-k) = m_{t-k} - B_{t-k}[a_{t-k+1} - a_t(-k+1)],$$
$$R_t(-k) = C^*_{t-k}\phi^{-1} - B_{t-k}[R^*_{t-k+1}\phi^{-1} - R^*_t(-k+1)\phi^{-1}]B'_{t-k},$$
$$B_t = C^*_t\phi^{-1}G'_{t+1}R^*_{t-1}\phi^{-1}.$$

The unconditional filtered distribution is derived as usual from

$$p(\theta_{t-k}|D_t) \propto \int p(\theta_{t-k}|D_t, \phi)p(\phi|D_t)\mathrm{d}\phi,$$

giving the t distribution

$$\theta_{t-k}|D_t \sim t_{n_t}[a_t(-k), (S_t/S_{t-k})R_t(-k)].$$

Notice that the scale parameter is adjusted by a factor determined by what is learned about the observation scale from the data following time $(t-k)$. Fitted values of the series are the filtered estimates of the mean response function, $\mu_t = F'_t\theta_t$, which are given by

$$\mu_{t-k} \sim t_{n_t}[f_t(-k), T_t(-k)],$$

where

$$f_t(-k) = F'_{t-k}a_t(-k),$$
$$T_t(-k) = (S_t/S_{t-k})F'_{t-k}R_t(-k)F_{t-k}.$$

In the case of no evolution noise ($W_t \equiv 0$) the filtered mean response is exactly equal to the ordinary regression fitted value.

Smoothing with Variance Discounting

When the observation scale is discounted, filtering is a little more problematic. First of all we require the set of filtered distributions,

$$\phi_{t-k}|D_t;$$

then the filtered state and mean response distributions are calculated as before. Difficulties arise, however, in the calculation of the filtered state distribution. Recall that no formal model is specified for scale evolution, the scale prior being directly defined as a discounted version of the previous posterior. Without a formal model it is not possible to use Bayes' theorem to filter information back from time t to time $(t-1)$ and beyond. Instead, an approximate method is used, the method of linear Bayes estimation. The joint distribution for ϕ_t and ϕ_{t-1} has moments

$$\begin{pmatrix} \phi_t \\ \phi_{t-1} \end{pmatrix} \bigg| D_{t-1} \sim \left[\begin{pmatrix} S_{t-1}^{-1} \\ S_{t-1}^{-1} \end{pmatrix}, \frac{2}{n_{t-1}S_{t-1}^2} \begin{pmatrix} \delta_t^{-1} & 1 \\ 1 & 1 \end{pmatrix} \right]$$

(the covariance being determined from an assumed linear evolution model, $\phi_t = \phi_{t-1} + u_t$ where u_t has mean 0 and variance $(\delta^{-1}-1)V[\phi_{t-1}|D_{t-1}]$, but an unknown distribution). Proceeding as in the derivation of filtered state distributions in the known variance case, but working now solely in terms of first and second order moments, yields expressions for the filtered mean and variance for ϕ_{t-1}. Assuming a gamma distribution for the filtered scale these moments uniquely define the parameters so that

$$\phi_{t-1}|D_t \sim G[n_t(-1)/2, d_t(-1)/2],$$

where

$$n_t(-1) = n_{t-1} + \delta_t(n_t(0) - \delta_t n_{t-1}),$$
$$S_t^{-1}(-1) = S_{t-1}^{-1} + \delta_t(S_t^{-1}(0) - S_{t-1}^{-1}),$$
$$d_t(-1) = n_t(-1)S_t(-1),$$

and starting values are $n_t(0) = n_t$, $S_t(0) = S_t$. Table 3.2 gives the general recurrence forms. West and Harrison (1989), Chapter 4, discusses linear Bayes estimation for the DLM in more depth, and Chapter 10 therein details the general analysis of stochastic variance modelling.

3.8 Forecast Monitoring

Assessing forecast performance provides information on model adequacy and indications of possible changes in data series structure. Performance is measured relative to the performance of alternative models, there being no sensible meaning to absolute measures of individual model performance. Alternative models are designed to be sensitive to particular departures from the routine model. Outlying observations, level shifts, and increased variation are oft-encountered general types of non-routine behaviour for which to watch out.

3.8.1 Bayes Factors

The model assessment tool is the Bayes factor, the relative predictive likelihood for two models. For example, consider the routine model M and an

3.8 Forecast Monitoring

Table 3.2 Summary of unknown variance model

Univariate DLM: unknown variance $V_t = k_t\phi^{-1}$, k_t **known**

Observation:	$Y_t = F_t'\theta_t + \nu_t \qquad \nu_t \sim N[0, k_t\phi_t^{-1}]$
System:	$\theta_t = G_t\theta_{t-1} + \omega_t \qquad \omega_t \sim t_{n_{t-1}}[0, W_t]$
Information:	$(\theta_{t-1} \mid D_{t-1}) \sim t_{n_{t-1}}[m_{t-1}, C_{t-1}]$
	$(\theta_t \mid D_{t-1}) \sim t_{\delta_t n_{t-1}}[a_t, R_t]$
	$(\phi_{t-1} \mid D_{t-1}) \sim G[n_{t-1}/2, d_{t-1}/2]$
	$(\phi_t \mid D_{t-1}) \sim G[\delta_t n_{t-1}/2, \delta_t d_{t-1}/2]$
	$a_t = G_t m_{t-1}$
	$R_t = G_t C_{t-1} G_t' + W_t$
Forecast:	$(Y_t \mid D_{t-1}) \sim t_{\delta_t n_{t-1}}[f_t, Q_t]$
	$f_t = F_t' a_t$
	$Q_t = F_t' R_t F_t + k_t S_{t-1}$

Updating Recurrence Relationships
$(\theta_t \mid D_t) \sim t_{n_t}[m_t, C_t]$
$(\phi_t \mid D_t) \sim G[n_t/2, d_t/2]$
$m_t = a_t + A_t e_t$
$C_t = (S_t/S_{t-1})[R_t - A_t A_t' Q_t]$
$e_t = Y_t - f_t \qquad\qquad A_t = R_t F_t/Q_t$
$n_t = \delta_t n_{t-1} + 1 \qquad d_t = \delta_t d_{t-1} + S_{t-1} e_t^2/Q_t$
$S_t = d_t/n_t$

Forecast Distributions

For $k \geq 1$,
$$(\theta_{t+k} \mid D_t) \sim t_{\delta_t n_t}[a_t(k), R_t(k)]$$
$$(Y_{t+k} \mid D_t) \sim t_{\delta_t n_t}[f_t(k), Q_t(k)]$$
$$a_t(k) = G_{t+k} a_t(k-1)$$
$$R_t(k) = G_{t+k} R_t(k-1) G_{t+k}' + W_{t+k}$$
$$f_t(k) = F_{t+k}' a_t(k)$$
$$Q_t(k) = F_{t+k}' R_t(k) F_{t+k} + k_{t+k} S_t$$
$$a_t(0) = m_t \qquad\qquad R_t(0) = C_t$$

Filtered Distributions

For $1 \leq k \leq t$,
$$(\theta_{t-k} \mid D_t) \sim t_{n_t(-k)}[a_t(-k), R_t(-k)]$$
$$(\phi_{t-k} \mid D_t) \sim G[n_t(-k)/2, d_t(-k)/2]$$
$$a_t(-k) = m_{t-k} - B_{t-k}[a_{t-k+1} - a_t(-k+1)]$$
$$R_t(-k) = C_{t-k} - B_{t-k}[R_{t-k+1} - R_t(-k+1)]B_{t-k}'$$
$$B_t = C_t G_{t+1}' R_{t+1}^{-1}$$
$$n_t(-k) = n_{t-k} + \delta_{t-k+1}(n_t(-k+1) - \delta_{t-k+1} n_{t-k})$$
$$S_t^{-1}(-k) = S_{t-k}^{-1} + \delta_{t-k+1}(S_t^{-1}(-k+1) - S_{t-k}^{-1})$$
$$d_t(-k) = n_t(-k) S_t(-k)$$

alternative model M_A with one step forecast distributions $p(Y_t|D_{t-1}, M)$ and $p(Y_t|D_{t-1}, M_A)$ respectively. The Bayes factor for the routine model against the alternative model is defined to be the ratio of these forecast distributions evaluated at the observed value

$$H_t = \frac{p(y_t|D_{t-1}, M)}{p(y_t|D_{t-1}, M_A)}.$$

If the observation $Y_t = y_t$ has a higher predictive density value under the routine model than under the alternative model, the Bayes factor exceeds one, and so on. Jeffreys (1961) suggests that a Bayes factor of more than 10 indicates evidence in favour of the routine model; more than 100 indicating strong evidence. In the other direction a Bayes factor of less than 1/10 indicates evidence in favour of the alternative model; less than 1/100 indicating strong evidence. Obviously a Bayes factor of 1 indicates no evidence either way, while values between 1/10 and 10 provide only slight evidence one way or the other.

3.8.2 Automatic Monitoring

Bayes factors for model comparison are used in an automatic forecast monitoring scheme in the following way. Given the routine model forecast, define a set of alternative forecasts to represent model breakdowns of interest, say a level increase or a variance increase. When the observation is made, compute the Bayes factors for the routine model against each of the alternatives. If any of the Bayes factors show evidence against the routine model, signal an exception. If there is no clear evidence against the routine model, it gets the benefit of the doubt and no signal is generated.

Such a monitoring scheme is clearly very flexible. Any specific kind of break with the routine model can be explicitly watched for, and the sensitivity of the monitor—the amount of evidence necessary against the routine model to signal an exception—can be adjusted at will.

3.8.3 Cumulative Evidence

The individual Bayes factor H_t measures the evidence for alternative models in the single observation at time t. Looking at groups of observations together is another way of assessing forecast performance. A Bayes factor for a group of observations—a cumulative Bayes factor—provides the required evidential tool. The cumulative Bayes factor for a consecutive set of observations $Y_t, Y_{t-1}, ..., Y_{t-k+1}$ is defined as the ratio of the joint forecast densities for those observations. It is equal to the product of the individual

3.8 Forecast Monitoring

Bayes factors

$$H_t(k) = \prod_{i=0}^{k-1} H_{t-i}$$

$$= \frac{p(Y_t|Y_{t-1},...,Y_{t-k+1},D_{t-k},M)\cdots p(Y_{t-k+1}|D_{t-k},M)}{p(Y_t|Y_{t-1},...,Y_{t-k+1},D_{t-k},M_A)\cdots p(Y_{t-k+1}|D_{t-k},M_A)}$$

$$= \frac{p(Y_t,Y_{t-1},...,Y_{t-k+1}|D_{t-k},M)}{p(Y_t,Y_{t-1},...,Y_{t-k+1}|D_{t-k},M_A)}.$$

Monitoring groups of observations in this way can point up evidence of change from the routine model that is slow to show itself. Time to time changes may never be individually large enough to provide sufficient doubt in the routine model to justify an exception, but taken together several similar reinforcing pieces of evidence can do so. A typical scenario is where a series exhibits a change in level—say an increase—but of a relatively small magnitude: small enough such that inherent observation variability can mask low forecasts when examined individually. But when a run of low forecasts occurs together, the cumulative Bayes factor monitor will notice the pattern and issue a signal.

3.8.4 Monitor Design

Designing forecast monitors involves making decisions on the twin issues of what kinds of forecast-observation discrepancies to look out for, and how much evidence of change is required before it is necessary to act. Both matters are highly application specific, but some general purpose suggestions are given in West and Harrison (1989), Chapter 11. Responding to exception signals is also a highly specific matter. Automatic response rules can be defined in particular situations and these will naturally be related to the properties of the signal and the weight of evidence in favour of particular departures from the routine model. In most cases signals will be handled on an individual basis, the forecaster responding after investigating possible causes of forecast failure.

3.8.5 General Monitoring Scheme

In a dynamic context it is important for a monitoring system to maintain special vigilance on very recent behaviour, and not to be influenced by judgements of past performance. Judgement that a model was satisfactory six months ago should not be allowed to bias evaluation of the model's performance this month.

When evaluating single observation performance with the Bayes factors H_t there is implicitly no such historical influence in the judgement. Only the current observation and model forecast are considered. But evidence of change collected over a series of observations through cumulative Bayes factors $H_t(k)$ does suffer from historical bias. If a model performs well for several periods, then evidence in its favour accumulates in a large cumulative Bayes factor. If a change subsequently occurs in the series, unless it is of sufficient magnitude to trigger the single observation judgement, it will be a long time before the existing accumulated favourable evidence becomes negated by accumulated unfavourable evidence. This is clearly not an efficient way to look out for trouble. The solution is to localise the judgement similar to the way in which the single observation judgement is localised. Examine the aggregate performance of the l most recent observations, discarding any influence from further into the past.

The general scheme is to reset the monitor to a neutral state of no evidence either way whenever the model is judged to have performed adequately. Evidence against the model will then accumulate directly from the time when forecast performance begins to deteriorate and a signal will be generated as soon as the evidential threshold is breached. No time will be lost in overcoming a stock of previously established good will.

The scheme is detailed in West and Harrison (1989) as follows. Define

$$L_t = H_t \times \min(1, L_{t-1}),$$

$$l_t = \begin{cases} 1 + l_{t-1} & L_{t-1} < 1, \\ 1 & L_{t-1} \geq 1, \end{cases}$$

initialised with $L_0 = 1$ and $l_0 = 1$. Then the sequence of *local* Bayes factors L_t provides the basis of a local monitoring scheme, with the *run length* l_t indicating how long ago the trouble may have begun.

If there is evidence in favour of the routine model at time $(t-1)$, the model is deemed adequate and future judgements will ignore the past. At the next time, t, the local Bayes factor is set to the Bayes factor for just that time, $L_t = H_t$. If the Bayes factor is very small, then the observation Y_t is a possible outlier or may indicate a structural change. If the Bayes factor exceeds one, then the model is adequate and the system rolls forward once more as before. Finally, if the Bayes factor lies between these two extremes, then there is some evidence against the model, but not enough to trigger an exception, and the evidence is held over for combined judgement with later evidence.

If there is inconclusive evidence against the routine model (the final scenario in the previous paragraph) before time t, so that the local Bayes

3.8 Forecast Monitoring

factor is less than one, $L_{t-1} < 1$, then evidence from the current observation is cumulated through the Bayes factor product $L_t = H_t \times L_{t-1}$, and the run length increases by one, $l_t = l_{t-1} + 1$. If now L_t is very small, then the l_t most recent observations together suggest evidence of reduced forecast performance. If L_t exceeds one, then, as above, previous doubts are assuaged and the model is deemed adequate. If there is still no decision either way, the evidence is held over once again.

An example will clarify the setup. Begin with a series of forecasts in good accord with the observed values.

Time $t = 1$

$$H_1 = 1.2 \Rightarrow L_1 = H_1 \times \min(1, L_0) = H_1 = 1.2,$$
$$l_1 = 1.$$

Time $t = 2$

$$H_2 = 1.3 \Rightarrow L_1 = 1.3 \times \min(1, 1.2) = 1.3 = H_2,$$
$$l_2 = 1.$$

Notice how the favourable evidence in the first forecast is excluded from the assessment of the second. While forecasts are 'good', the local Bayes factor just equals the Bayes factor for the current observation and the run length is one. And naturally no signals are generated.

Now observe some poorer forecasts (τ denotes the assigned evidence limit— Bayes factor threshold—below which a signal is generated).

Time $t = 3$

$$0.3 = \tau < H_3 = 0.7 < 1 \Rightarrow L_3 = 0.7 \times \min(1, 1.3) = 0.7 = H_3,$$
$$l_3 = 1.$$

There is some evidence of model inadequacy, $L_3 < 1$, but it is not conclusive, $L_3 > \tau$. Hold over the evidence to the next time.

Time $t = 4$

$$\tau < H_4 = 0.4 < 1 \Rightarrow L_4 = 0.4 \times \min(1, 0.7) = 0.28,$$
$$l_4 = 2.$$

The monitor signals ($L_4 < \tau$) with evidence accumulated over two periods, a run length of two. At this point the forecaster, alerted to a potential problem, will investigate possible causes of the perceived breakdown in forecast performance and possibly intervene to adjust the model description before continuing. The monitor is reset to a neutral state before proceeding.

Time $t = 5$

$$\tau < H_5 = 0.9 < 1 \Rightarrow L_5 = 0.9 \times \min(1,1) = 0.9,$$
$$l_5 = 1.$$

Notice that in the calculation of the local Bayes factor the previous value is reset from 0.28 to 1. The evidence that led to the signal at time 4 is removed from the picture—because it has already been brought to the forecasters attention and appropriate action taken. Just as a build-up of favourable historical performance is prevented from giving bias to current judgements, so is historically unfavourable evidence once it has been acted upon.

Time $t = 6$

$$\tau < H_5 = 0.7 < 1 \Rightarrow L_6 = 0.7 \times \min(1, 0.9) = 0.63,$$
$$l_6 = 1 + 1 = 2.$$

Evidence is once again building up against the routine model, $L_6 < 1$. It is not yet sufficient to cause a signal ($L_t > \tau$) so it is held over.

Time $t = 7$

$$H_7 = 2 \Rightarrow L_7 = 2 \times \min(1, 0.63) = 1.26,$$
$$l_7 = 2 + 1 = 3.$$

An observation in very good accord with the model, $L_7 > 1$, removes the concern that had been building, the model is judged satisfactory, and so the evaluation next time will start from a neutral position.

Time $t = 8$

$$H_8 = 0.8 \Rightarrow L_8 = 0.8 \times \min(1, 1.26) = 0.8,$$
$$l_8 = 1.$$

Time 8 begins from a new evidential origin as the favourable position from time 7 is neutralised by the factor $\min(1,1.26)$.

One final scenario remains. It is possible for a series of observations to consistently provide evidence against the model (exemplified in times 3 and 4 above), but in sufficiently small quantities that the run length grows considerably without the Bayes factor threshold being breached. One may want to be made aware of this without having to wait too long. Such awareness is accomplished by extending the monitor to issue a signal whenever the run length exceeds a preset limit in addition to signalling when the Bayes factor threshold is passed.

3.9 Error Analysis

One of the most important aspects of any stastical modelling is the examination of model residuals. Residuals are what is left unexplained after our current efforts to understand and represent data. The reader will be familiar with the importance and techniques of residual analysis from the study of 'ordinary' regression models. In the time series setting, matters are not much different, except we have the distinction between forecast and fitted residuals.

In non-time-series regression models and also in classical time series modelling (see Appendix 3.2) residual analysis is confined to the set of residuals from a model fitted to a data set. However, in the forecasting context attention is more usefully focussed on examination of forecast residuals. In the DLM one step ahead forecast residuals are independent and distributed according to the Student-t distribution on n_{t-1} degrees of freedom (or the normal distribution when the observation variance is known). For degrees of freedom greater than about 15 the normal distribution provides a good approximation for residual analysis purposes.

In examining forecast residuals one is looking for evidence that the model is lacking in some way: unusually outlying points, clusters of similarly signed residuals, clusters of greater or lesser variability, and any other kind of structure. The discussion of forecast monitoring in the previous section is relevant here: forecast monitoring is a form of on-line residual analysis. For examining residuals a simple time plot is indispensable. Histograms, normal quantile plots, and autocorrelations are also extremely useful.

Rather than discuss the details further here we refer you to discussions in the case studies in Chapters 4-6 where residual analysis can be seen in the context of practical model building, time series analysis, and forecasting. General discussion can be found in West and Harrison (1989), Chapter 10, and Box and Jenkins (1976), Chapter 8.

3.10 References

Box, G.E.P. and Jenkins, G.M. (1976). *Time Series Analysis: forecasting and control.* Holden-Day, San Francisco.

Broemeling, L.D. (1985). *Bayesian Analysis of Linear Models.* Marcel Dekker, New York.

Jeffreys, H. (1961). *Theory of Probability.* Oxford University Press, London.

Pole, A. and West, M. (1990). Efficient Bayesian learning in nonlinear dynamic models. *J. Forecasting*, 9, 119-136.

Stevens, C.F. (1974). On the variability of demand for families of items. *Oper. Res. Quart.*, **25**, 411–420.

Smith, A.F.M. and West, M. (1983). Monitoring renal transplants: an application of the multi-process Kalman filter. *Biometrics*, **39**, 867–878.

West, M. and Harrison, P.J. (1989). *Bayesian Forecasting and Dynamic Models*, Springer-Verlag, New York.

3.11 Exercises

Exercise 3.1 For the static simple random walk model (zero evolution variance) with known observation variance $V_t = V$, show that the mean and variance of the k step ahead forecast, $Y_{t+k}|D_t$, are $Q_t(k) = C_t + V$ and $f_t(k) = m_t$ respectively.

Exercise 3.2 Refer to Subsection 3.2.5. Derive the result in the text for the moments of $p(\theta_t|D_t)$ where $(\theta_{t-2}|D_{t-2}) \sim N[m_{t-2}, C_{t-2}]$ and the sum $Z_t = Y_t + Y_{t-1}$ is observed but the individual values for Y_t and Y_{t-1} are not.

Exercise 3.3 *Continuation.* For the constant trend (random walk) model show that the derived posterior $p(\theta_t|D_t)$ for block updating is the same as the posterior that results from the sequential updating with each observation individually when the individual series values are known. Try to prove the result for general models.

Exercise 3.4 Constant DLM. Your company sells an ethical pharmaceutical drug Kurit which currently sells about an average of 100 units each month. As from January ($t = 1$) a new formulation with new packaging but the same name will replace the existing product. Initial beliefs about the new underlying demand μ are that the most likely value is 130 units and the uncertainty is described by a standard deviation of 20 units with $(\mu_0|D_0) \sim N[130, 400]$. For forecasting purposes it is decided to use the first order polynomial constant DLM:

$$\text{Observation Equation}: \quad Y_t = \mu_t + \nu_t, \quad \nu_t \sim N[0, V],$$
$$\text{System Equation}: \quad \mu_t = \mu_{t-1} + \omega_t, \quad \omega_t \sim N[0, W],$$

with $V = 100$, $W = 5$, and $(\mu_0|D_0) \sim N[130, 400]$. Assuming no further inputs of market/subjective information:

(a) Sequentially produce forecasts for January to October.
(b) Plot the adaptive factor (A_t) against t. What do you conclude?
(c) Roughly sketch the forecast distributions for $t = 1$ and $t = 10$.

3.11 Exercises

Here is the calculation sequence:
  (i) Level forecast variance, $R_t = C_{t-1} + W$.
  (ii) Forecast mean, $f_t = m_{t-1}$, and variance, $Q_t = R_t + V$.
  (iii) Adaptive factor, $A_t = R_t/Q_t$.
  (iv) Forecast error, $e_t = y_t - f_t$.
  (v) Level posterior mean, $m_t = m_{t-1} + A_t e_t$, and variance, $C_t = A_t V$.

The sales actually recorded are, from January to September, 150, 136, 143, 154, 135, 148, 128, 149, 146.

Repeat the exercise with the alternative prior settings:
  (i) Very vague: $(\mu_0|D_0) \sim N[500, 250^2]$.
  (ii) Very sure: $(\mu_0|D_0) \sim N[130, 1]$.

Note: It is recommended that you do this exercise by hand to see how the information in the observation series feeds into the model components. If you want to try variations with BATS, a constant variance is approximated by setting a very high number for the initial degrees of freedom, 10,000 for example, and variance discount factor 1.

Exercise 3.5 Component discounting. The setup is the same as in the previous exercise. In this case we use component discounting instead of constant variance evolution noise. The only change to the model is that now $W_t = C_{t-1}(\delta^{-1} - 1)$. For this exercise use $\delta = 0.8$. The calculation sequence is:
  (i) Level forecast variance, $R_t = C_{t-1}/\delta$.
  (ii) Forecast mean, $f_t = m_{t-1}$, and variance, $Q_t = R_t + V$.
  (iii) Adaptive factor, $A_t = R_t/Q_t$.
  (iv) Forecast error, $e_t = y_t - f_t$.
  (v) Level posterior mean, $m_t = m_{t-1} + A_t e_t$, and variance, $C_t = A_t V$.

Try the same initial conditions as in the previous exercise.

Exercise 3.6 Variance learning. The previous exercise is now extended to include learning on the observation variance, V, previously assumed constant and known. The model observation and system equations are unchanged; however, the variance learning introduces (a) change of marginal distribution of components from normal to Student t, $(\mu_t|D_t) \sim t_{n_t}[m_t, C_t]$, and (b) variance estimation quantities $(n_t S_t/V|D_t) \sim \chi^2_{n_t}$, where S_t is the estimate of V (on n_t degrees of freedom). For the exercise $m_0 = 130$, $C_0 = 400$, $\delta = 0.8$, $n_1 = 1$, $d_1 = 100$, so that $S_1 = 100$. The sequence of calculations is:
  (i) Level forecast variance, $R_t = C_{t-1}/\delta$.
  (ii) Forecast mean, $f_t = m_{t-1}$, and variance, $Q_t = R_t + S_{t-1}$.
  (iii) Adaptive factor, $A_t = R_t/Q_t$.
  (iv) Forecast error, $e_t = y_t - f_t$.

(v) Variance 'sum of squares', $d_t = d_{t-1} + S_{t-1} e_t^2 / Q_t$, degrees of freedom, $n_t = n_{t-1} + 1$, estimate, $S_t = d_t / n_t$.

(vi) Level posterior mean, $m_t = m_{t-1} + A_t e_t$, and variance, $C_t = A_t S_t$.

Compare results with the previous exercise: what do you conclude? What is the forecast for October, $(Y_{10} | D_9)$?

Exercise 3.7 Intervention. In September ($t = 9$) news is received that, because of suspected toxic side effects, the major competitive product Burnit, which has 50% of the product market, is to be withdrawn from the market at the end of the month until further notice. All Burnit patients will have to transfer to other brands and the effect, $\Delta \mu_{10}$, of this on Kurit's October sales is uncertain. It is assessed that with roughly a 2/3 probability the effect will be to increase sales by between 100 and 180, thus approximately doubling sales. This information is formulated as $\Delta \mu_{10} \sim N[140, 800]$. The model is the first order polynomial constant discount DLM with system intervention:

Observation Equation : $Y_t = \mu_t + \nu_t,$ $\nu_t \sim N[0, V],$
System Equation : $\mu_t = \mu_{t-1} + \omega_t + \Delta \mu_t,$ $\omega_t \sim N[0, W_t].$
Subjective Information : $\Delta \mu_t \sim N[\Delta m_t, \Delta R_t],$

For this exercise continue from Exercise 3.5 with $\delta = 0.8$ and known observation variance $V = 100$. In March ($t = 15$) news is broken that Burnit is cleared of toxic effects and is to be reintroduced to the market in April ($t = 16$). However, it is thought that Kurit will retain some of the former Burnit patients and that the drop in sales will be roughly between 40 and 120, formulated as $\Delta \mu_{10} \sim N[-80, 1600]$. (Other than at $t = 10$ and $t = 16$ there is no external change, $\Delta \mu_t = 0$.) The calculation sequence is:

(i) Level forecast variance, $R_t = C_{t-1}/\delta + \Delta R_t$.
(ii) Forecast mean, $f_t = m_{t-1} + \Delta m_t$, and variance, $Q_t = R_t + V$.
(iii) Adaptive factor, $A_t = R_t / Q_t$.
(iv) Forecast error, $e_t = y_t - f_t$.
(v) Level posterior mean, $m_t = m_{t-1} + \Delta m_t + A_t e_t$, variance, $C_t = A_t V$.

The observations for October to the following August are 326, 292, 327, 301, 315, 304, 187, 169, 192, 175, 188.

Exercise 3.8 Simple dynamic regression. I is a lead indicator for a company's sales S. A simple discount DLM relates Y_t, the quarterly change in S_t, to F_t, the quarterly change in I_{t-2}. The (known observation variance) constant discount regression model is:

Observation Equation : $Y_t = F_t \theta_t + \nu_t,$ $\nu_t \sim N[0, V],$
System Equation : $\theta_t = \theta_{t-1} + \omega_t,$ $\omega_t \sim N[0, W_t].$

3.11 Exercises

Use constant observation variance $V = 1$, component discount $\delta = 0.8$, and initial information $m_0 = 2$, $C_0 = 0.81$. The relevant distributions are:

(i) Regression coefficient prior: $\theta_t | D_{t-1} \sim N[m_{t-1}, R_t]$.
(ii) Forecast: $Y_t | D_{t-1} \sim N[f_t, Q_t]$.
(iii) Regression coefficient posterior: $\theta_t | D_t \sim N[m_t, C_t]$.

The calculation sequence is:

(i) Regression coefficient forecast variance, $R_t = C_{t-1}/\delta$.
(ii) Forecast mean, $f_t = F_t m_{t-1}$, and variance, $Q_t = F_t^2 R_t + V$.
(iii) Adaptive factor, $A_t = F_t R_t / Q_t$.
(iv) Forecast error, $e_t = y_t - f_t$.
(v) Regression coefficient posterior mean, $m_t = m_{t-1} + A_t e_t$, and variance, $C_t = R_t V / Q_t$.

The regression and observation series values, Y_t and F_t, are $(12, 4)$, $(11, 4)$, $(9, 3)$, $(5, 2)$, $(3, 1)$, $(0, -1)$, $(-7, -3)$, $(-7, -4)$, $(-6, -3)$, $(-3, -1)$, $(7, 2)$, $(10, 3)$, $(13, 4)$, $(12, 4)$. Plot the following: (i) the forecasts, f_t, against the outcomes, y_t, (ii) the regression coeffcent posterior estimate, m_t, against time, t, and (iii) the regression coefficient estimate, m_t, against the regression series, F_t. Comment on what you observe.

# Appendix 3.1

Review of Distribution Theory

In this appendix we review the essential distribution theory that underlies the statistical analysis of the univariate dynamic linear model. We present definitions, properties, and results without proof. A more extensive discussion is given in West and Harrison (1989), Chapter 16. Detailed discussions of distribution theory in Bayesian analysis can be found in DeGroot (1970), Box and Tiao (1973), and Aitchison and Dunsmore (1976). Readers interested in a comprehensive treatment of distribution theory more generally should consult the excellent multi-volume reference by Johnson and Kotz (1970, 1971, 1972).

Univariate Normal Distribution

A random quantity X is said to have a *normal distribution* or to be *normally distributed* with mean μ and variance σ^2 if it has the probability density function

$$p(X) = \frac{1}{\sqrt{2\pi}\sigma} \exp\left[-\frac{1}{2}\left(\frac{x-\mu}{\sigma}\right)^2\right], \qquad (\sigma > 0).$$

We use the notation $X \sim N[\mu, \sigma^2]$.

Sums of Normal Variables

Let X_i, $i = 1, ..., k$ be k independent normally distributed variables with means μ_i and variances σ_i^2 respectively, $X_i \sim N[\mu_i, \sigma_i^2]$. Define a new variable Z to be the weighted sum $Z = \sum_i^k a_i X_i$. Then Z is also normally distributed. The mean and variance of Z are weighted sums of the means and variances respectively of the component X_i:

$$E[Z] = \sum_{i=1}^{k} a_i E[X_i] = \sum_{i=1}^{k} a_i \mu_i,$$

$$V[Z] = \sum_{i=1}^{k} a_i^2 V[X_i] = \sum_{i=1}^{k} a_i^2 \sigma_i^2.$$

More generally, let the X_i variables be correlated with pairwise covariances $C[X_i, X_j] = \sigma_{ij}$ for $i \neq j$. Z is still normally distributed with the mean given previously; the variance, however, is now extended to include contributions from the covariances among the X_i,

$$V[Z] = \sum_{i=1}^{k} a_i^2 V[X_i] + 2 \sum_{i=1}^{k} \sum_{j=1}^{i-1} a_i a_j C[X_i, X_j]$$

$$= \sum_{i=1}^{k} a_i^2 \sigma_i^2 + 2 \sum_{i=1}^{k} \sum_{j=1}^{i-1} a_i a_j \sigma_{ij}.$$

Multivariate Normal Distribution

A random p-vector X is said to be (jointly) normally distributed with mean vector μ and covariance matrix Σ if it has the (joint) probability density function

$$p(X) = \frac{1}{(2\pi)^{p/2} |\Sigma|^{p/2}} \exp\left[\frac{1}{2}(x - \mu)' \Sigma^{-1}(x - \mu)\right].$$

An equivalent definition is: a set of p random quantities X_i is jointly normally distributed if and only if every linear combination, $\sum_{i=1}^{p} a_i X_i$ with at least one $a_i \neq 0$, is normally distributed. We use the notation $X \sim N[\mu, \Sigma]$.

Normal variables are independent if and only if they are uncorrelated. If Σ is diagonal, then the elements X_i are mutually independent.

Linear Transformations

Let L be a $(q \times p)$ matrix of constants (with $q \leq p$) and define a new variable Z to be the linear map $Z = LX$. Then Z is normally distributed with mean $L\mu$ and covariance matrix $L\Sigma L'$, $Z \sim N[L\mu, L\Sigma L']$.

Let $X_i \sim N[\mu_i, \Sigma_i]$ for $i = 1, ..., k$ be independent normal p-vectors. Define $Z = \sum_{i=1}^{k} A_i X_i$ for constant $(l \times p)$ matrices A_i with $l \leq p$. Then Z is normally distributed with mean $\sum_{i=1}^{k} A_i \mu_i$ and covariance matrix $\sum_{i=1}^{k} A_i \Sigma_i A_i'$. Notice how similar this result is to the corresponding univariate result, the similarity being reflected in the notation.

The result on summation of normals is easily seen to be a special case of the general linear transformation. The linear algebra is straightforward. Pile the individual X_i on top of each other to form a single kp-vector. Define L to be the $(p \times pk)$ transformation matrix $A_1 I_p ... A_k I_p$ where I_p is the $(p \times p)$ identity matrix. In the context of the normal distribution the piling operation is permissible because normal variables can always be regarded as the margins (see next section) of a larger multivariate normal vector (recall the alternative definition given above).

Marginal Distributions

Let X be a normally distributed random p-vector, $X \sim N[\mu, \Sigma]$. Let X be partitioned as $X' = (X_a', X_b')$ where X_i are p_i-vectors $(i = a, b)$ and $p = p_a + p_b$. Define conformable partitions of the mean vector and covariance matrix,

$$\mu = \begin{pmatrix} \mu_a \\ \mu_b \end{pmatrix}, \qquad \Sigma = \begin{pmatrix} \Sigma_{aa} & \Sigma_{ab} \\ \Sigma_{ba} & \Sigma_{bb} \end{pmatrix}.$$

Then the partitions are (marginally) distributed as normal, $X_i \sim N[\mu_i, \Sigma_i]$, $i = a, b$. When $p_i = 1$ the univariate X_i is univariate normally distributed. An alternative way to see these results is through the general result for linear transformations of normal variables, presented in the previous subsection. For example, if $L = (0, ..., 0, 1, 0, ...0)$ where the 1 appears in the jth position, then $Z = LX = X_j$.

Conditional Distributions

Partition X as in the previous subsection. Then conditioning on X_b the distribution of X_a is normal, $X_a | X_b \sim N[\mu_{a|b}, \Sigma_{a|b}]$, with the mean and covariance given by

$$\mu_{a|b} = \mu_a + \Sigma_{ab}\Sigma_{bb}^{-1}(X_b - \mu_b),$$
$$\Sigma_{a|b} = \Sigma_{aa} - \Sigma_{ab}\Sigma_{bb}^{-1}\Sigma_{ba}.$$

The matrix $A_a = \Sigma_{ab}\Sigma_{bb}^{-1}$ is called the *regression matrix* of X_a on X_b. The conditional moments may be expressed in terms of the regression matrix as

$$\mu_{a|b} = \mu_a + A_a(X_b - \mu_b),$$
$$\Sigma_{a|b} = \Sigma_{aa} - A_a\Sigma_{bb}^{-1}A_a',$$

a form that is used in the DLM updating equations. The vertical bar in $X_a|X_b$, for example, denotes conditioning.

Gamma Distribution

A positive random quantity ϕ is said to have a *gamma distribution* with parameters $n > 0$ and $d > 0$ if it has the probability density function

$$p(\phi) = \frac{d^n}{\Gamma(n)}\phi^{n-1}\exp(-\phi d),$$

where Γ is the gamma function. The mean and variance are

$$E[\phi] = n/d,$$
$$V[\phi] = E[\phi]^2/n = n/d^2.$$

We use the notation $\phi \sim G[n, d]$.

Univariate Student-t Distribution

A random quantity X is said to have a Student-t distribution on n degrees of freedom with mode μ and scale parameter $\tau > 0$ if it has the probability density function

$$p(X) = \frac{\Gamma[(n+1)/2]n^{n/2}}{\Gamma(n/2)(\pi\tau)^{1/2}}\left[n + \frac{(x-\mu)^2}{\tau}\right]^{-(n+1)/2}.$$

The mean is $E[X] = \mu$ and the variance is $V[X] = n\tau/(n-2)$ if $n > 2$. We use the notation $X \sim t_n[\mu, \tau]$.

Multivariate Student-t Distribution

A random p-vector X is said to have a joint Student-t distribution on n degrees of freedom with mode μ and scale matrix Ω if it has the (joint) probability density function

$$p(X) = \frac{\Gamma[(n+p)/2]n^{(n+p-1)/2}}{\Gamma(n/2)\pi^{p/2}|\Omega|^{p/2}} \left[n + (x-\mu)'\Omega^{-1}(x-\mu)\right]^{-(n+p)/2}.$$

We use the notation $X \sim t_n[\mu, \Omega]$.

Marginal and conditional distributions for subvectors of X parallel those given for the multivariate normal distribution. A result that is of particular interest here is that the univariate margin for an individual element of X has a univariate Student-t distribution on n degrees of freedom with mode μ_i and scale parameter Ω_{ii}.

Joint Normal-Gamma Distributions

The combination of the normal distribution (for means) and gamma distribution (for precision or inverse variance) occurs often in Bayesian analysis of linear models, including analysis of the DLM, because of their convenient mathematical properties.

Univariate Normal-Gamma

Let X be a conditionally normally distributed random quantity so that $X|\phi \sim N[m, C\phi^{-1}]$, where m and C are known constants. Let ϕ be a gamma random quantity, $\phi \sim G[n/2, d/2]$ for any $n > 0$ and $d > 0$. The joint distribution of X and ϕ is called the (univariate) normal-gamma distribution.

The joint probability density function of X and ϕ is just the product $p(X|\phi)p(\phi)$ from which it is easily deduced that the conditional distribution $\phi|X$ has density

$$p(\phi|X) \propto \left(\frac{\phi}{2\pi C}\right)^{1/2} \exp\left[-\frac{\phi(x-m)^2}{2C}\right] \frac{d^{n/2}}{2^{n/2}\Gamma(\frac{n}{2})} \phi^{\frac{n}{2}-1} \exp\left[-\frac{\phi d}{2}\right]$$

$$\propto \phi^{\frac{n+1}{2}-1} \exp\left[-\frac{\phi}{2}\left\{\frac{(x-m)^2}{C} + d\right\}\right].$$

This is the form of a gamma distribution with parameters $(n+1)/2$ and $(C^{-1}(x-m)^2 + d)/2$.

The density function of the marginal distribution of X is similarly easily deduced by integrating over ϕ in the joint density:

$$p(X) = \int_0^\infty p(X|\phi)p(\phi)\mathrm{d}\phi$$

$$\propto \left[n + \frac{(x-m)^2}{R}\right]^{-(n+1)/2},$$

where $R = Cd/n = C/E[\phi]$. This is proportional to the density of the Student-t distribution with n degrees of freedom, mode m, and scale parameter R.

Multivariate Normal-Gamma

Let the random p-vector X be conditionally normally distributed so that $X|\phi \sim N[m, C\phi^{-1}]$ where the p-vector m and $(p \times p)$ symmetric positive definite matrix C are known. Here, each element of the covariance matrix of X is scaled by the common factor ϕ. Let ϕ be a gamma random quantity, $\phi \sim G[n/2, d/2]$ for any $n > 0$ and $d > 0$. The results given in the univariate X case now generalise to the multivariate X case in an obvious manner.

The conditional distribution of $\phi|X$ is gamma with parameters $(n+p)/2$ and $[d + (x-m)'C^{-1}(x-m)]/2$. The 'degrees of freedom' increases by the dimension of X, namely p; the 'sum of squares' increases by the quadratic form generalisation of the univariate case. Obviously, when $p = 1$ these results reduce to the univariate results given in the previous section.

The marginal distribution of X is the multivariate analogue of the univariate result, that is, the p dimensional multivariate Student-t distribution with n degrees of freedom, mode m, and scale matrix $R = C(d/n) = C/E[\phi]$.

References

Aitchison, J. and Dunsmore, I.R. (1975). *Statistical Prediction Analysis*, Cambridge University Press, Cambridge.

Box G.E.P. and Tiao G.C. (1973). *Bayesian Inference in Statistical Analysis*. Addison-Wesley, Reading, MA.

DeGroot M.H. (1970). *Optimal Statistical Decisions*. McGraw-Hill, New York.

Johnson N.L. and Kotz, S. (1970). *Distributions in Statistics: Continuous Univariate Distributions – I*, Wiley, New York.

References

Johnson N.L. and Kotz, S. (1971). *Distributions in Statistics: Continuous Univariate Distributions – II*, Wiley, New York.

Johnson N.L. and Kotz, S. (1972). *Distributions in Statistics: Continuous Multivariate Distributions*, Wiley, New York.

West, M. and Harrison, P.J. (1989). *Bayesian Forecasting and Dynamic Models*, Springer-Verlag, New York.

Appendix 3.2

Classical Time Series Models

In this appendix we briefly describe the class of autoregressive moving average (ARMA) models and show how they are related to the dynamic linear model.

Autoregressive Models

Autoregressive models relate the current value of a series, y_t, to values the series exhibited in the past, $\{y_{t-k} : k = 1, 2, ...\}$. The simplest case is the first order autoregressive model, abbreviated as AR(1), which is defined by

$$y_t = \phi y_{t-1} + \epsilon_t,$$

where $\{\epsilon_t\}$ is a series of independent, zero mean, constant variance, normally distributed stochastic variables. More generally, the observation series may have a nonzero mean, μ, and values of the series further into the past than the immediately preceding time period may be included in the dependency. The general autoregressive model of order p, denoted AR(p), is defined by

$$y_t - \mu = \phi_1(y_{t-1} - \mu) + \cdots + \phi_p(y_{t-p} - \mu) + \epsilon_t, \qquad \epsilon_t \sim N[0, \sigma^2].$$

Notice that the model parameters, the autoregressive coefficients, ϕ_i, and the series mean, μ, do not vary with time.

Moving Average Models

Moving average models relate the current value of the observation series to current and past unobservable stochastic variables. The first order moving average, or MA(1), model is defined as

$$y_t - \mu = \epsilon_t + \psi_1 \epsilon_{t-1},$$

where $\{\epsilon_t\}$ is a series of independent, zero mean, constant variance, normally distributed stochastic variables.

Autoregressive Moving Average Models

Combining a pth order autoregression and a qth order moving average yields the autoregressive moving average model, ARMA(p,q),

$$y_t - \mu = \sum_{i=1}^{p} \phi_i (y_{t-i} - \mu) + \sum_{j=1}^{q} \psi_j \epsilon_{t-j} + \epsilon_t.$$

The ARMA class of models is set out in detail in the classic work of Box and Jenkins (1976). The mathematical structure and statistical analysis of ARMA models is detailed therein and a comprehensive modelling methodology is developed. Other, more recent, treatments are listed in the references.

DLM Representation of ARMA Models

The traditional way in which ARMA models can be represented in the structural form of the DLM is as follows. Define $m = \max(p, q+1)$, $\phi_j = 0$ for $j = p+1, ..., m$ and $\psi_k = 0$ for $k = q+1, ..., m$, then

$$y_t = \sum_{i=1}^{m} (\phi_i y_{t-i} + \psi_i \epsilon_{t-i}) + \epsilon_t.$$

Now define a DLM through

$$F = \begin{pmatrix} 1 \\ 0 \\ \vdots \\ 0 \\ 0 \end{pmatrix}, \quad G = \begin{pmatrix} \phi_1 & 1 & 0 & \cdots & 0 \\ \phi_2 & 0 & 1 & \cdots & 0 \\ \vdots & \vdots & \vdots & \ddots & \vdots \\ \phi_{m-1} & 0 & 0 & \cdots & 1 \\ \phi_m & 0 & 0 & \cdots & 0 \end{pmatrix}, \quad \omega_t = \begin{pmatrix} 1 \\ \psi_1 \\ \vdots \\ \psi_{m-2} \\ \psi_{m-1} \end{pmatrix} \epsilon_t,$$

and $\nu_t \equiv 0$. The structural form here will be familiar to engineers as the Kalman filter model.

DLM Representation of ARMA Models

Example: MA(1)

As an example consider the MA(1) model. Here $p = 0$ and $q = 1$ so that $m = 2$ and the DLM components are

$$F = \begin{pmatrix} 1 \\ 0 \end{pmatrix}, \quad G = \begin{pmatrix} 0 & 1 \\ 0 & 0 \end{pmatrix}, \quad \omega_t = \begin{pmatrix} 1 \\ \psi_1 \end{pmatrix} \epsilon_t.$$

Expanding the system equation $\theta_t = G\theta_{t-1} + \omega_t$ it is easy to see that

$$\theta_{2,t} = \psi_1 \epsilon_t,$$
$$\theta_{1,t} = \theta_{2,t-1} + \epsilon_t$$
$$= \psi_1 \epsilon_{t-1} + \epsilon_t.$$

Finally the observation is equated with the first element of the state vector, $y_t = F'\theta_t$.

Example: AR(1)

For the first order autoregressive model, $p = 1$ and $q = 0$ so that $m = 1$ and the DLM representation is

$$F = 1, \quad G = \phi_1, \quad \omega_t = \epsilon_t.$$

Expanding the system equation yields the infinite sum

$$y_t = \epsilon_t + \phi_1 \epsilon_{t-1} + \phi_1^2 \epsilon_{t-2} + \cdots,$$

which is precisely the infinite moving average form of the AR(1) model.

Alternative Representations of AR Models

Autoregressive models may be expressed in an alternative (perhaps more intuitive) form, namely

$$F_t = \begin{pmatrix} y_{t-1} \\ \vdots \\ y_{t-p} \end{pmatrix}, \quad \theta_t = \begin{pmatrix} \phi_1 \\ \vdots \\ \phi_p \end{pmatrix}, \quad G = I_p, \quad \omega_t = 0, \text{ and } \nu_t = \epsilon_t.$$

In this form it is easier to observe that forecasting more than one step ahead requires future values of the series being forecasted. The need for knowledge of future values of series is a feature of all regression component DLMs when predicting step ahead. Practical approaches to the problem are discussed in Chapter 5.

Another often used representation is given by

$$F = \begin{pmatrix} 1 \\ 0 \\ 0 \\ \vdots \\ 0 \end{pmatrix}, \quad G = \begin{pmatrix} \phi_1 & \phi_2 & \cdots & \phi_p \\ 1 & 0 & \cdots & 0 \\ 0 & 1 & \cdots & 0 \\ \vdots & \vdots & \ddots & \vdots \\ 0 & 0 & \cdots & 1 \end{pmatrix}, \quad \omega_t = \begin{pmatrix} \epsilon_t \\ 0 \\ 0 \\ \vdots \\ 0 \end{pmatrix}, \quad \text{and } \nu_t \equiv 0.$$

Stationarity

Stationarity is a property required by a series in order to apply classical ARMA model analysis. In the formulation of autoregressive models in the preceding section a nonzero system evolution innovation variance, W_t, results in a nonstationary model. In the general formulation given above, ARMA models in DLM form become nonstationary if a nonzero observation variance is added.

Definition *A time series* $\{y_t : t = 1, 2, ...\}$ *is said to be* stationary *if the joint distribution of any collection of k values is invariant with respect to arbitrary shifts of the time axis.*

In terms of joint probability densities stationarity may be expressed thus: for integers $k \geq 1$ and $s \geq 0$,

$$p(y_{t_1}, ..., y_{t_k}) = p(y_{t_1+s}, ..., y_{t_k+s}).$$

Definition *A time series* $\{y_t : t = 1, 2, ...\}$ *is said to be* weakly stationary *or second order stationary if the mean, variance, and covariances are time-invariant. For integers* $t > 0$ *and* $s < t$,

$$E[y_t] = \mu,$$
$$V[y_t] = \sigma^2,$$
$$\text{Cov}[y_{t-s}, y_t] = \gamma_s.$$

The requirement of stationarity imposes strict conditions on the possible values of ARMA model parameters. From the infinite moving average representation of the AR(1) model exemplified above it is clear that a finite variance requires that the autoregressive coefficient be less than one in magnitude, $|\phi_1| < 1$. This is equivalent to saying that the root of the *characteristic polynomial*,

$$\phi(m) = 1 - \phi_1 m = 0,$$

must be greater than one in absolute value. The general condition for ARMA models is that all of the roots of the characteristic polynomial,

$$\phi(m) = 1 - \phi_1 m - \cdots - \phi_p m^p = 0,$$

are greater than one in absolute value, and that all of the roots of the characteristic polynomial,

$$\theta(m) = 1 + \theta_1 m + \cdots + \theta_q m^q = 0,$$

are less than one in absolute value. See the references for further discussion.

Modelling with ARMA Models

Most time series are nonstationary. A first step in practical modelling with ARMA models is to transform the observation series into as close as possible a stationary series. Transformations to stabilise variance, familiar from standard regression analysis, are supplemented in time series analysis with time dependent transformations called *differencing*. A time series exhibiting a constant drift in trend may be transformed to a stationary series (no mean drift) by taking first differences, $Z_t = Y_t - Y_{t-1}$. The *back shift* operator, B, is used to denote differencing, $Z_t = (1 - B)Y_t$. Higher order differences remove polynomial trends; for example, the second order (or twice repeated) difference $W_t = (1 - B)^2 Y_t = Y_t - 2Y_{t-1} + Y_{t-2}$ removes a constantly growing drift in trend. When a series is differenced in this way the original undifferenced series is called *integrated*. An ARMA model applied to a differenced series is referred to as an ARIMA model for the original series. ARIMA stands for autoregressive integrated moving average. As an example, an ARMA(0,2) model applied to a first differenced series is an ARIMA(0,1,2) model for the raw series.

Seasonality can removed by seasonal differencing. For monthly data exhibiting an annual cycle the twelfth *seasonal difference* $(1 - B^{12})Y_t$ removes the seasonality. Of course, such differencing removes any linear drift in mean also.

Quite apart from the use of differencing transformations to induce stationarity in a time series, plots of differenced series (stationary or not to begin with) can be a useful exploratory device. Unusual data points can be dramatically highlighted in a difference plot. Chapter 7 includes an example.

Definition *The* autocovariance at lag s *of a time series* $\{y_t : t = 1, 2...\}$ *is the covariance of the series with itself lagged by s periods. For integers $t > 0$ and $s < t$,*

$$\gamma_s = \text{Cov}[y_{t-s}, y_t].$$

Notice that for a stationary series the autocovariance function is time-invariant, depending only upon the lag between values and not the actual timing. That is, $\mathrm{Cov}[y_{t-s}, y_t] = \mathrm{Cov}[y_{t-s+k}, y_{t+k}]$ for any integer $k \geq 0$.

Definition *The autocorrelation at lag s of a time series $\{y_t : t = 1, 2...\}$ is the variance scaled autocovariance*

$$\rho_s = \frac{\gamma_s}{\sigma^2}$$

where $\sigma^2 = \mathrm{Var}[y_t]$.

Note that the instantaneous autocovariance is just the series variance, and the corresponding autocorrelation is unity, $\gamma_0 = \sigma^2$ and $\rho_0 = 1$.

Once a stationary series is obtained by differencing the autocorrelation structure may be modelled within the ARMA class. Box and Jenkins (1976) shows how, including how to construct forecasts for the original observation series.

The strategy for reducing a time series to stationarity is often useful but is only workable if the nonstationarities addressed by differencing are static. If they are not, the difference transformed series will be less nonstationary (perhaps) than the original, but it will definitely not be stationary. Dynamic models obviate the stationarity problem and may be more useful.

Forecast Function Equivalence of ARMA and DLM

The first order polynomial DLM—the steady model—is detailed in Section 3.5. Particularising the DLM updating recurrence relationships (Table 3.1) to this model we have

$$e_t = Y_t - m_{t-1},$$
$$m_t = m_{t-1} + A_t e_t.$$

We may therefore write

$$Y_t - Y_{t-1} = e_t - (1 - A_t) e_{t-1}.$$

It is a simple exercise to show that, for the model with fixed observation and evolution variances, $V_t = V$ and $W_t = W$, the so-called constant model, the adaptive factor A_t has limiting form

$$\lim_{t \to \infty} A_t = A = \frac{\sqrt{1 + 4V/W}}{2V/W},$$

and the one step forecast variance Q_t has limiting form

$$\lim_{t \to \infty} Q_t = Q = V/(1 - A).$$

Therefore, taking the limit as time proceeds to infinity yields

$$\lim_{t \to \infty}[Y_t - Y_{t-1} - e_t + (1 - A)e_{t-1}] = 0.$$

We therefore have the limiting model

$$Y_t = Y_{t-1} + \epsilon_t + \psi\epsilon_{t-1},$$

where $\psi = -(1 - A)$ and the quantities ϵ_t are independently normally distributed, $\epsilon_t \sim N[0, Q]$. Thus, the limiting form of the forecast function of the first order polynomial DLM is that of an ARIMA(0,1,1) process. (Note that in the limiting form the moving average terms are in fact the one step ahead forecast errors.)

In like manner it can be seen that the form of the limiting forecast function of the constant variance second order polynomial trend DLM (the linear growth model) is that of an ARIMA(0,2,2) process. More general results are given in West and Harrison (1989), Chapters 5 and 7.

References

Aoki, M. (1990). *State Space Modelling of Time Series.* Springer-Verlag, New York.

Box, G.E.P. and Jenkins, G.M. (1976). *Time Series Analysis: forecasting and control.* Holden-Day, San Francisco.

Harvey, A.C. (1989). *Forecasting, Structural Time Series Models, and the Kalman Filter*, Cambridge University Press, Cambridge.

Pankratz, A. (1991). *Forecasting with Dynamic Regression Models.* Wiley, New York.

Pole, A. and West, M. (1990). Efficient Bayesian learning in non-linear dynamic models, *Journal of Forecasting*, **9**, 119–136.

Shumway, R.H. (1988). *Applied Statistical Time Series Analysis*, Prentice-Hall, Englewood Cliffs, NJ.

Stuart, A. and Ord, K. (1993). *Kendall's Advanced Theory of Statistics*, Volume 3, Griffin, London.

Vandaele, W. (1983). *Applied Time Series and Box-Jenkins Models*, Academic Press, New York.

Wei, W.W.S. (1990). *Time Series Analysis.* Addison-Wesley, Reading, MA..

West, M. and Harrison, P.J. (1989). *Bayesian Forecasting and Dynamic Models.* Springer-Verlag, New York.

Chapter 4

Application: Turkey Chick Sales

In this and the next three chapters we apply the models and methods discussed in the previous chapters to time series from several disparate fields of study. This chapter and Chapters 5 and 6 develop individual detailed case studies. Chapter 7 describes preliminary analyses for a large number of additional series. Each of the detailed applications is designed to develop and illustrate particular aspects of times series analysis and forecasting, though there are, of course, features common to all. The first application (this chapter) derives from commerce and is concerned with forecasting transformed sales series. The second application (Chapter 5), again from commerce, is concerned with forecasting market share and assessing the importance of possible explanatory factors. The third application (Chapter 6) is an examination of sociological data, an investigation of the numbers of marriages in Greece. In contrast to the two previous applications this study focuses on time series analysis rather than forecasting. Points of special interest are a long term periodic effect and comparison of regional patterns.

Chapter 7 contains a series of short descriptions of additional data sets with suggestions for possible analyses. The series are drawn from areas including environmental studies (water quality analysis, zooplankton/nutrient analysis, zooplankton/phytoplankton analysis), commerce (housing starts and sales, advertising outlay and sales, cigarette consumption), sociology (numbers of live births, numbers of U.K. marriages), energy consumption, finance (exchange rates), and medicine. All of the data sets are included on the accompanying program diskette.

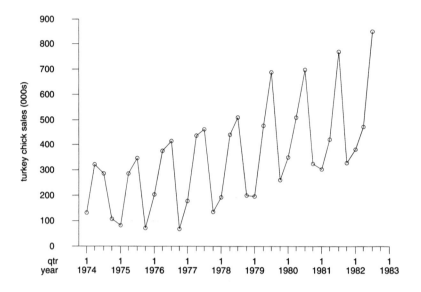

Figure 4.1 Quarterly turkey chick sales in Eire.

4.1 Preliminary Investigation

In this application we explore forecasting for a nonlinear time series. Figure 4.1 shows the quarterly total sales (in thousands) of one day old turkey chicks from hatcheries in Eire (Ameen and Harrison, 1985a). The series exhibits sustained growth over the recorded period 1974 to 1982, together with a pronounced seasonal pattern. The seasonal pattern shows a definite increase in the peak to trough variation (seasonal amplitude) as the level increases.

4.1.1 Stabilising Variation

Small unsystematic changes in seasonal amplitude are well modelled by the dynamic seasonal DLM with a seasonal discount factor less than unity. The turkey chick series, however, has more exploitable structure than such an analysis would assume. The amplitude increase quite definitely appears to have a direct relationship with the underlying series level. Modelling such structure explicitly should provide better forecasts than assuming the changes to be purely random. Determining whether the structure is real and not simply apparent is accomplished by examining forecast performance for alternative models representing structure versus no structure.

4.1 Preliminary Investigation

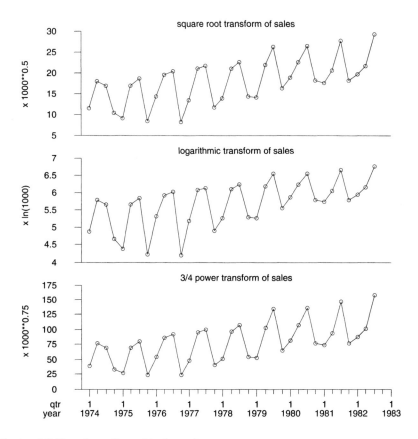

Figure 4.2 Transformations of turkey sales.

The technique of data transformation is one useful way in which data structures of the amplitude-trend relationship kind may be directly addressed. Natural candidates for variance stabilisation in the present case are the logarithmic transformation and power transformations. Figure 4.2 shows time plots of three representative transformations of the turkey data: the natural logarithmic transform, the square root transform, and the 3/4 power transform. Each of these transformations of the original raw series removes the increasing amplitude effect as you can see from the figure. But the transformations do not all perform equally well in terms of removing the nonlinearity.

The logarithm series, in particular, presents even more of a problem for modelling than the raw untransformed series: Notice how the seasonal pattern over the first four years has a much greater amplitude—and in fact increasing over the years—than thereafter.

The square root and 3/4 power transformed series are much better behaved. Of the two, the 3/4 power series appears to exhibit the most consistent overall behaviour and is therefore the series most likely to be best forecast. However, the differences apparent between these two series are not very great and visual impressions can be misleading. It behoves us to examine both series in more detail—to build a model and compare forecast performance for each. Moreover, other power transformations should also be considered—there is a whole range of values between the raw series, power transformation index 1.0, and the square root series, power transformation index 0.5—and we invite you to do so in the exercises.

4.1.2 Seasonal Pattern Changes

There is another interesting feature of the seasonal pattern easily discernible by visual inspection of the time plot in Figure 4.1. The form of the pattern—its shape—undergoes a clear change after 1978. The basic form of seasonal variation up to the end of 1978 can be described as 'low, high, high, low'. From 1979 onwards the pattern changes to 'lower, low, high, lower': a change from a square shape to a triangular shape.

An important question to answer is whether the change is a gradual evolution or the result of a sudden structural shift in the turkey chick market. It is interesting to address this question informally while comparing the raw series and the power transformed series. On the original scale the change in 1979 appears to be well demarcated, suggesting a sudden shift. On the transformed scale, however, the picture is much less clear-cut. The seasonal pattern changes could very likely be the the result of smooth dynamic variation over several years. Inferences drawn from analysis of the data must be careful in assessing these effects. It is the raw sales series that holds our primary interest, not an 'artificial' series utilised for modelling expediency.

4.1.3 Forecasting Transformed Series

A model comprised of a linear growth trend and an unrestricted seasonal pattern was selected for the turkey data. We remarked earlier on the sustained growth pattern and clearly a dynamic constant level is inappropriate for that. A restricted seasonal pattern (for quarterly data that means either the first or the second harmonic only) was considered but proved to be inadequate. This inadequacy was expected given the seasonal pattern changes visually apparent in the data.

4.1 Preliminary Investigation

Table 4.1 Component prior information

Component	Raw Mean (Sd.)	Square Root Mean (Sd.)	3/4 power Mean (Sd.)
Level	225(20)	15(0.75)	58(4.0)
Growth	0(10)	0(0.3)	0(2.0)
Seasonal Effect 1	−100(10)	−4(0.5)	−21(2.5)
Seasonal Effect 2	100(10)	4(0.5)	21(2.5)
Seasonal Effect 3	100(10)	4(0.5)	21(2.5)
Seasonal Effect 4	−100(10)	−4(0.5)	−21(2.5)
Observation sd	5 on 1 dof	1 on 1 dof	2 on 1 dof

The model was investigated using a range of discount factors for the trend (0.9–1.0) and seasonal (0.6–1.0) components. Both square root and 3/4 power transformed series were analysed with the result that the latter series proved to be the better of the two scalings on which to forecast.

Prior information was assessed by examining turkey sales for a few years preceding the data set (pre-1974). Component priors were determined to be as given in Table 4.1; one characteristic to note about these priors is the relatively large degree of uncertainty.

The prior estimates on the transformed scales (columns three and four in Table 4.1) are obtained from the priors for the raw sales series to give approximately equivalent initial information on the square root and 3/4 power scales. For example, taking an upper value of the mean plus two standard deviations, 265, gives 16.3 on the square root scale and 65.7 on the 3/4 power transform scale. With transformed means of 15 and 58 on the latter scales respectively we use the upper limits as approximate two standard deviation limits and set the standard deviations to 0.75 and 4.

To begin the investigation we examine analyses using no forecast monitoring. In practice one would typically always implement a monitoring scheme since potential problems that would be flagged are otherwise being ignored, but a monitor-free analysis is a useful first exercise for comparing the alternative transformations, and a base from which to demonstrate the benefits of more interactive analysis or management by exception.

4.1.4 Assessing Transformations

The family of power transformations, parameterised by power index λ, has observation equation
$$Y_t^\lambda = F_t'\theta_t + \nu_t.$$
It is a straightforward matter to show that the likelihood for the power

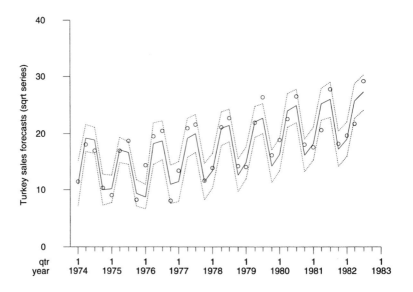

Figure 4.3 One step forecasts: square root transformation, static model.

index is given by the likelihood for the raw observations series multiplied by the Jacobian $J = \lambda^n \prod_{i=1}^{n} y_i^{\lambda-1}$ (Box and Cox, 1963).

The raw log-likelihood statistic (which is the figure reported by the BATS program) is therefore adjusted by adding the logarithm of the Jacobian, $n \log \lambda + (\lambda - 1) \sum_{i=1}^{n} \log y_i$, so that comparisons of alternative power indices may be made. (Assuming a uniform prior on λ these adjusted loglikelihoods are log-posterior values for λ.)

Square Root Transformation

Table 4.2 reports summary forecast statistics for a representative set of discount factors for the trend and seasonal components.

On the basis of log-posterior probabilities the best forecasting performance is achieved with a trend discount of about 0.9 and seasonal discount of about 0.8. This is an atypical situation since most often the seasonal development of a series is more stable than the trend. The classical measures of point forecast performance, mean squared error and mean absolute deviation, yield a similar judgement. By comparison with the dynamic model the static model (both discount factors equal to unity) is substantially inferior on all three measures. Figures 4.3 and 4.4 graphically illustrate the differences.

4.1 Preliminary Investigation

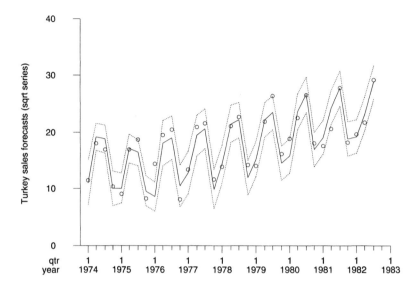

Figure 4.4 One step forecasts: square root transformation, $\delta_T = 0.9$, $\delta_S = 0.8$.

Table 4.2 Analysis summary statistics for square root transform

Discounts	MSE	MAD	Log Likelihood
$\delta_T = 0.90$, $\delta_S = 0.8$	3.11	1.34	−71.1
$\delta_T = 0.95$, $\delta_S = 0.8$	3.22	1.35	−71.7
$\delta_T = 1.00$, $\delta_S = 1.0$	4.23	1.64	−77.6

The changing seasonal pattern is only very slowly adapted to by the static model. (Remember that as information accrues, on-line estimates change even with a static model: we are *not* considering model fit here.) By the third quarter of 1983 the estimated pattern is only very slightly different to the starting square pattern.

The dynamic model does rather better. The seasonal pattern evolves from the starting square shape to the finishing triangular shape quite well, the improved forecast performance over the static model being readily discernible visually.

Forecasting beyond the data provides another view of the comparative performance of the static and dynamic models. Table 4.3 shows the point

Table 4.3 Prediction summary

Time	Static Model Mean	Sd
1982 IV	18	3.3
1983 I	20	3.3
1983 II	26	3.3
1983 III	29	3.3

Dynamic Model: $\delta_T = 0.9$, $\delta_S = 0.8$		
Time	Mean	Sd
1982 IV	19	2.8
1983 I	21	2.7
1983 II	23	2.6
1983 III	30	2.5

forecasts and 90% uncertainty measures for one year following the data set. The uncertainties from the static model are from one sixth to one quarter larger than the corresponding uncertainties from the dynamic model. The initial prior square seasonal pattern is strongly evident (the forecasts are for 1982-IV to 1983-III). The forecasts from the dynamic model by contrast reflect the triangular pattern prevalent in the latter part of the data set. In the absence of any information relating to further changes of form the more recent cyclical behavioural pattern is more believable as a projection of future behaviour.

3/4 Power Transformation

Analyses of the 3/4 power transformed data were performed with the same grid of trend and seasonal discounts used in the analyses of the square root transformed data. Table 4.4 contains a summary of representative results. The general pattern of summary forecast statistics tells a story similar to the square root analyses. Overall a trend and seasonal discount pair around $(0.9, 0.7)$ gives the best forecast performance, with the static model faring quite poorly by comparison.

Figures 4.5 and 4.6 show the one step forecasts for the static model and the best dynamic model. Comparison reveals features analogous to the comparison of the corresponding models for the square root series (in Figures 4.4 and 4.5).

4.1 Preliminary Investigation

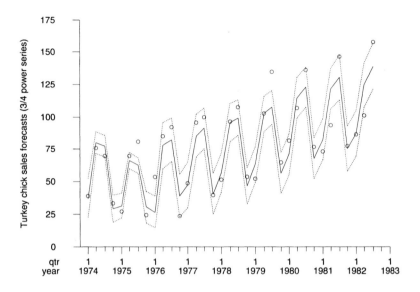

Figure 4.5 One step forecasts: 3/4 power transformation, static model.

Table 4.4 Analysis summary statistics for 3/4 power transform

Discounts	MSE	MAD	Log-Likelihood
$\delta_T = 0.90$, $\delta_S = 0.7$	111.0	7.9	-134.7
$\delta_T = 0.95$, $\delta_S = 0.7$	114.2	7.9	-135.3
$\delta_T = 1.00$, $\delta_S = 1.0$	153.6	9.6	-144.2

Choosing a Transformation

The Jacobians of the square root and 3/4 power transformations (see the technical outline at the beginning of this section) are evaluated to be

$$J_{1/2} = 35 \log(1/2) + (1/2 - 1) \sum_{i=1}^{n} \log y_i$$
$$= -123.66,$$

$$J_{3/4} = 35 \log(3/4) + (3/4 - 1) \sum_{i=1}^{n} \log y_i$$
$$= -59.77.$$

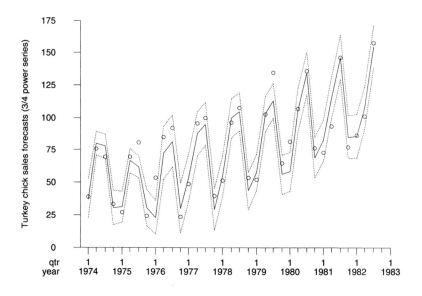

Figure 4.6 One step forecasts: 3/4 power transformation, $\delta_T = 0.9$, $\delta_S = 0.7$.

Table 4.5 Prediction summary

	Static Model	
Time	Mean	Sd
1982 IV	83	19
1983 I	94	19
1983 II	129	19
1983 III	150	19
Dynamic Model: $\delta_T = 0.9$, $\delta_S = 0.7$		
Time	Mean	Sd
1982 IV	87	15
1983 I	94	14
1983 II	110	13
1983 III	164	12

4.1 Preliminary Investigation

Table 4.6 Adjusted log-likelihoods

	Square Root Transformation: $\lambda = 1/2$	
Discounts	Log-Likelihood	Adjusted Log-Likelihood
(0.90, 0.8)	−71.1	−194.8
(0.95, 0.8)	−71.7	−195.4
(1.00, 1.0)	−77.6	−201.3
	3/4 Power Transformation: $\lambda = 3/4$	
Discounts	Log-Likelihood	Adjusted Log-Likelihood
(0.90, 0.7)	−134.7	−194.5
(0.95, 0.7)	−135.3	−195.1
(1.00, 1.0)	−144.2	−204.0

Table 4.6 shows the adjusted log-likelihoods (log-posterior probabilities assuming a uniform prior) corresponding to the values recorded in Tables 4.2 and 4.4. The (log) posterior probabilities are the joint probabilities for the power transformation index and component discount factors. There is not a great deal to choose between the two transformations, but certainly the 3/4 power transformation has the edge over the square root transformation.

Forecast Comparison

Tables 4.3 and 4.5 present step ahead forecasts for one year following the data for each of the transformed series. In practical terms, regardless of the choice of modelling scale determined by variance stabilisation or other considerations, forecasts will be required on the raw data scale. The mean and variance of probability distributions do not transform directly (except for linear transformations) but percentage points do. The square root forecast for 1982-IV has mean (which is also the median for symmetric distributions like the normal and t) 18 and 90% limits $(18 - 3.3)$ and $(18 + 3.3)$ which translate to 18^2, $(18-3.3)^2$, and $(18+3.3)^2$ for the median and 90% limits on the raw scale.

Notice that the uncertainty limits on the raw scale are not symmetric. Squaring symmetrically distributed variables does not produce a variable with a symmetric distribution; rather it produces a skewed distribution (and this is, of course, why moments do not directly transform). Table 4.7 shows the step ahead forecasts on the raw scale for the dynamic models selected as best for the square root and 3/4 power transformations. (See also Tables 4.3 and 4.5; note that in all tables rounding is done after calculation so that, for example, squaring the means in Table 4.3 does not necessarily yield the mode reported in Table 4.7.)

Table 4.7 Predictions on observation scale

Period	Square Root Transformation: $\lambda = 1/2$	
	Forecast Mode	90% interval
1982 IV	380	(278, 498)
1983 I	422	(317, 507)
1983 II	547	(431, 676)
1983 III	910	(763, 1069)

Period	3/4 Power Transformation: $\lambda = 3/4$	
	Forecast Mode	90% Interval
1982 IV	382	(299, 469)
1983 I	424	(317, 507)
1983 II	525	(431, 676)
1983 III	898	(814, 983)

The point forecasts for both models are very close, but the 3/4 power transform model has consistently shorter uncertainty intervals.

4.2 Live Forecasting

The analyses reported above assume a single pair of component discount factors for trend and seasonal for all times. This implicitly assumes that although some movement is allowed by the dynamic formulation (both discounts are less than one) there are no unexpectedly large movements in the data. Those analyses serve the purposes of demonstrating that the data does require a time varying model quantification, and how certain kinds of nonlinearities may be modelled using transformations within the framework of dynamic linear models.

Practical forecasting goes further than such analysis allows. When forecasting time series live one continually assesses how well the forecasts are performing and checks the forecasts against eventual outcomes to keep a watch for indications of change. When possible changes are identified, forecasts may be altered to reflect information pertaining to the situation where such information is not already included in the formal modelling mechanism. This is management by exception. (Forecast monitoring systems are discussed in Section 2.5; technical details are given in Section 3.8.) Moreover, past data can be reanalysed once points of change have been more clearly identified, enabling better estimates of the effects of change. Information gleaned that way may be useful in making better informed judgements of future changes.

4.2 Live Forecasting

4.2.1 Forecast Analysis

The analysis of the previous section suggested that for a constant dynamic the trend and seasonal discount factors should be around 0.9 and 0.7 respectively. That kind of analysis is a compromise because any structural changes other than smooth dynamic variation pull component discount factors in a downward direction. While this does allow for the structural changes to be reasonably adapted to, it also means that during routine intervals there is excessive uncertainty injected during evolution resulting in imprecise forecasts.

When the forecast system is explicitly on the lookout for structural changes, such a compromise over discounting is not necessary. Higher discounts would typically be expected to give better forecast performance. This is indeed the case with the turkey sales series. Trend and seasonal discount factors of 0.95 and 0.8 were found to be more suitable (considering the same combinations as in the previous section) for the 3/4 power transformation series with no specific information available to make well informed interventions at monitor break points. We report analyses for this combination of discounts only in what follows. (You may find it helpful in following the commentary to mask Figure 4.11, revealing the data and forecasts as the commentary proceeds. Running the analysis through BATS is also beneficial.)

The model is initialised with the same prior specification as in the previous section, final column of Table 4.1. Forecasts for the first 18 months are very accurate. The monitor system first issues a signal in the third quarter of 1975 warning of an observation quite large in comparison with the forecast. In the absence of any specific information on the possible nature of a change here—Has the seasonal pattern changed? Has the market level increased for some reason? Are there special circumstances surrounding turkey sales this quarter?—the default action is to ignore the outlying value, increase component prior uncertainties substantially, and take careful notice of the next observation.

The observation for 1975-IV agrees with the forecast suggesting that the previous (large) value was not the start of a lasting change.

The observed value for the first quarter of 1976 is, like the value six months previously, significantly higher than the forecast and once again the monitoring system issues an alert. The outcome is, in fact, twice as large as the forecast, and considerably greater than has hitherto been seen for a first quarter figure. Lack of specific knowledge of any system changes once again leads to the default signal response: ignore the observed value, increase component uncertainties, and await further data before making a judgement.

Table 4.8 Implied and actual priors for 1976/3

Component	Model Prior	User Prior
Level	54 (15.4)	59(8)
Growth	0.14(2.3)	0(1)
Seasonal Effect 1	−22 (2.2)	−22(3)
Seasonal Effect 2	32 (2.2)	32(3)
Seasonal Effect 3	12 (2.2)	12(3)
Seasonal Effect 4	−22 (2.2)	−22(3)
Observation std. dev.	1.56	2

The forecast for the next quarter, 1976-II, is once again very low. The automatic monitoring system does not evaluate the discrepancy as significant to warrant a warning signal—because the forecast variance is quite large following the action taken in response to the signal in the previous quarter—however, from the forecaster's viewpoint there is something worth investigating here.

The pattern of the observations for the first half of 1976 certainly follows the pattern established for previous years: the second quarter being of the order of 60–80 units larger than the first. But both values are significantly greater than hitherto experienced. Market analysis reveals no reason to suspect a change in seasonal pattern, and little social or economic evidence for an increase in yearly sales level of the order of 50 units that is implied by the first six months figures if the current seasonal pattern is extrapolated. (The figures in the discussion here relate to the observation scale, not to the transformation scale on which the analysis is being performed. Recall the point we have made before: communication is always in terms of quantities directly meaningful to the investigator.) Judgement is made that a level increase of around 15 (real) units is justified. Component variances are already large: in fact they are considered too large and adjusted downward in spite of the real uncertainty over the level increase. Table 4.8 shows the priors for 1976-III (on the 3/4 power transformed scale we are using in the analysis) implied by the model based analysis and the modified priors actually used.

Note the form of the seasonal pattern here. It reflects the pattern *before* 1975 wherein the third quarter effect is smaller than the second. Recall that in the analysis thus far, the observation for 1975-III was ignored so there has been no information insofar as the analysis is concerned to change the view. However, our intervention on the level has resulted in an increase in the forecast for 1976-III over the forecast for 1976-II. The former is again considerably lower than the outcome and a monitor signal is issued.

4.2 Live Forecasting

Table 4.9 Implied and actual priors for 1976/4

Component	Model Prior	User Prior
Level	76(4.4)	70 (7)
Growth	0(1.3)	0 (1)
Seasonal Effect 1	−23(3.9)	−21.5(7)
Seasonal Effect 2	31(3.9)	21.5(7)
Seasonal Effect 3	15(3.9)	31.5(7)
Seasonal Effect 4	−23(3.9)	−31.5(7)
Observation std. dev.	1.97	Unchanged

The picture for 1976 now emerging is strongly supportive of both an increase in level to around 65-70 units (rather more than was thought just three months ago) and a change in seasonal pattern. Quarter three now appears to have a larger effect than quarter two. This latter inference is drawn from the current evidence for 1976 and also from the evidence of 1975. It begins to look as though 1975-III was the first occurrence of a changed seasonal pattern—though this was not known at the time and we ignored both the evidence and the observation in the analysis.

The problem confronting the forecaster looking forward to the final quarter of 1976 is whether the level increase is quite so substantial or whether there is going to be an offsetting low fourth quarter. Judgement is that the fourth quarter will follow the usual pattern of turning out to be around the first quarter value, but probably somewhat lower so that the 1976 level is not quite so substantially greater than the 1975 level as 15–20 units. The priors (on the 3/4 power transform scale) are adjusted as in Table 4.9.

The fourth quarter of 1976 turns out to be considerably below the level of the first quarter; the misgivings surrounding the level increase in 1976 have turned out to be well founded—overall there *has* been an increase, but not as much as the first nine months suggested by comparison with earlier years. The fourth quarter was indeed an offsetting factor—but note that it is no lower than the previous year's value for the fourth quarter. Forecast uncertainty is large enough that no monitor signal is generated.

Figure 4.7 shows the position at the end of 1976 and Figure 4.8 shows how the current view projects for 1977.

Looking back over market development from 1974, and considering current economic conditions, beliefs are that the new seasonal pattern is well established, around the right level, but that there is definite growth that is expected to continue. The prior on the growth is therefore adjusted upward from its present value of zero to two units. Other component priors are unchanged. Table 4.10 summarises the prior settings at the outset of 1977.

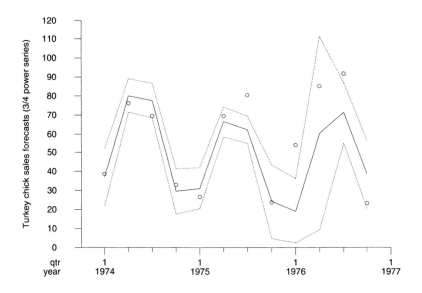

Figure 4.7 Position at end of 1976.

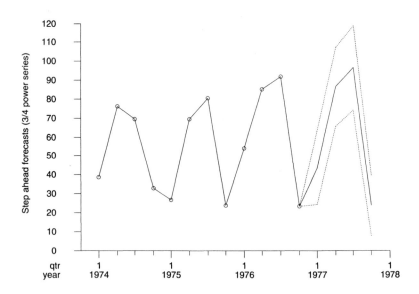

Figure 4.8 Forecasts for 1977.

4.2 Live Forecasting

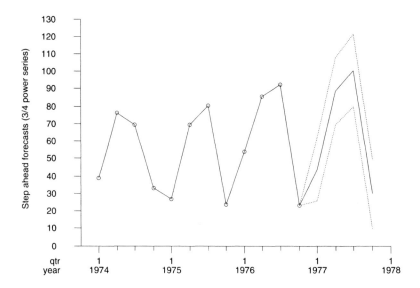

Figure 4.9 Post-intervention Forecasts for 1977.

Table 4.10 Implied and actual priors for 1977/1

Component	Model Prior	User Prior
Level	63(5.1)	Unchanged
Growth	0(1.1)	2(1)
Seasonal Effect 1	−19(8.1)	Unchanged
Seasonal Effect 2	24(8.1)	Unchanged
Seasonal Effect 3	34(8.1)	Unchanged
Seasonal Effect 4	−39(8.1)	Unchanged
Observation std. dev.	2.1	Unchanged

Figure 4.9 shows the forecast position for 1977 following this adjustment of the expected growth. Notice how, by comparison with the forecasts in Figure 4.8, the underlying level is projected to increase over the course of the year.

Forecasts for 1977 and 1978 are quite acceptable (notice how the uncertainty decreases with each observation in agreement with the forecast) and no adjustments are necessary. The same is true for the first half of 1979 but in 1979-III there is a very large number of sales. At the time there is no apparent reason for such a large increase over what was expected and the decision is taken to ignore the value in the analysis. Prior variances are increased to reflect additional uncertainty arising from the surprise value. Table 4.11 shows the existing and revised priors for the final quarter of 1979.

The outturn in 1979-IV is in good accord with the forecast, supporting the view just taken that the previous quarter's large sales value was a fluke. But the forecast for 1980-I is considerably lower than the outcome. Is this, like 1976-I, an indication of change (seasonal or otherwise)? There has just been a high quarter three value: the situation is very uncertain and there are no useful indications apart from the sales figures on which to base assessment. Judgement is to continue—including this high value in the analysis—the priors are reasonably uncertain so we will just keep a wary eye on developments.

The second quarter outcome is a little lower than the forecast; nothing to be concerned about, although there is the possibility that it is in some way an offsetting value to compensate for the previous high value. We will have to be open to the possibility of another seasonal pattern change. No action is taken on the priors for 1980-III, which turns out to be higher than the forecast. Notice that the seasonal development for the first nine months of 1980 is similar to that exhibited in 1979: a pattern rather different from that which prevailed over 1975–1978 (or prior to 1975). One important question that poses itself at this point is will there be a greater drop in the final quarter to offset the cumulatively high sales in the first nine months? No such drop-off was observed in 1979—when the new seasonal pattern was first established—so the decision is to go on without further change. The outcome for 1980-IV confirms the judgement call.

At the end of 1980 what is the outlook for 1981? There does seem to be a new seasonal pattern set in place, and market analysis sees no reason for this to change (as with all previous outlooks, a position of stability seems to prevail). However, the third quarters of the two preceding years have been very high; combined with the overall picture, this suggests the probability of a slow-down in growth. The current rate of 2.4(0.7) units is believed to be over optimistic and is lowered to 2.0(0.7). Figure 4.10 shows the forecasts for 1981 based on these revised priors.

The first quarter of 1981 turns out to be lower than expected, even after the growth estimate retrenchment. The second quarter is also low. Perhaps growth really has stopped following the large increases over the last few years. With these two low outturns in the analysis the growth estimate as we enter the third quarter is down to 1 with great uncertainty at 1.8; judgement is to proceed without intervention at this time. The third and fourth quarter forecasts are quite good; the previous judgement call seems to have been borne out. Moving into 1982 there are no known reasons for change and indeed the forecasts for the first nine months (to the end of the

4.3 Retrospective Perspectives

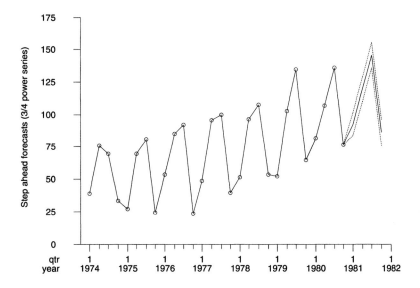

Figure 4.10 Forecasts for 1981.

Table 4.11 Implied and actual priors for 1979/4

Component	Model Prior	User Prior
Level	84 (2.2)	unchanged
Growth	1.5(0.4)	1.5(1)
Seasonal Effect 1	−26 (2.5)	−26 (4)
Seasonal Effect 2	22 (2.5)	22 (4)
Seasonal Effect 3	30 (2.5)	30 (4)
Seasonal Effect 4	−26 (2.5)	−26 (4)
Observation std. dev.	1.99	Unchanged

data set) are in very close agreement with the data. Figure 4.11 shows the complete live forecast analysis.

4.3 Retrospective Perspectives

One of the features of the live analysis of the previous section is the continual reassessment of unusual (past) observations in the light of later information. Two points in particular that stand out in this regard are the

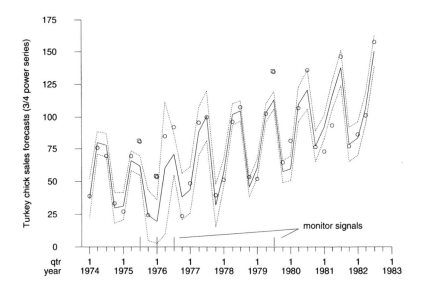

Figure 4.11 Forecasts for live analysis.

third quarters of 1975 and 1979. At the time sales in both periods were unusually large for no apparent reason and the figures were therefore discounted from the analysis; sometime later it was realised that the points were in fact the onset of an underlying change in the seasonal pattern for turkey sales and therefore important information had been discarded.

Forecasting positions were reassessed in the light of these realisations and suitable modifications made to component priors before continuing with the analysis and making further forecasts. At the time we did not elaborate on the details of how the judgements for the adjustments were determined, relying on phrases like 'based on the past two years it is judged that...'. That is a perfectly valid approach, but such assessments can be done more systematically using retrospective analysis of the model, taking a backward looking view.

4.3.1 The View From 1977

At the end of 1976 it became clear that the seasonal pattern underwent a slight change of form in 1975. At that time we could rework the analysis making a better informed decision of how to treat the third quarter of 1975: include the sales value for that time (instead of treating it as an outlier and omitting it) in the analysis because it is highly informative about changed

4.3 Retrospective Perspectives

dynamics, and assess the type and magnitude of the changes using the retrospective model analysis.

Make a forward intervention in the analysis at 1975-III to increase the component prior uncertainties so as to allow for adjustment to the change we know begins at that time. The level standard deviation is increased from 2 to 5; the growth estimate changed from negative, -2 with standard deviation 0.5, to positive, $2(2)$; and the seasonal factor standard deviations increased from 2 to 5. (The sales value for 1976-I is still regarded as unusually high from the standpoint of the end of that year, from whence we are taking this backward looking view, so it is still treated as an outlying point in this analysis.) Figure 4.12 shows how the forecast analysis now progresses up to the end of 1976. These are not forecasts that we could have made at the time because we did not then know the nature of the change in 1975-III. Of course, if we did have the benefit of that knowledge the improved forecasts shown here could have been realised.

Figures 4.13 and 4.14 show the fitted estimates and estimated seasonal pattern for the revised analysis. From the latter graph we obtain a clear picture of how the seasonal pattern has changed; it is this view of the turkey sales world that we now expect to propagate into the future.

Take a look once again at Figure 4.8. That figure shows the result of our reassessment of the implications of 1975/3 made at the end of 1976 without using the formal analysis here. The current analysis produces forecasts that are less uncertain, Figure 4.15. There are two reasons for the variance reduction. We preempted the onset of change and allowed the model to adapt earlier; forecast errors are therefore smaller which leads to greater certainty in estimates. Also, we did not include the additional interventions at 1976-III and 1976-IV, both of which increased uncertainty and were necessary in the live analysis because of the lack of knowledge of what was happening at the time. As you can see, reassessment of the past in the light of new information can produce real benefits even if done informally. Potentially greater gains, exemplified here in the reduction of uncertainty, are available with a formal analysis. Moreover, reassessment can be done at any time.

4.3.2 The Global Picture

Before leaving the turkey sales analysis we will take a look at what the series as a whole has to say, from the standpoint of the end of 1983 looking back. This analysis will give us a clear picture of the characteristics and magnitudes of the changes that actually occurred in the market during the period 1974–1983.

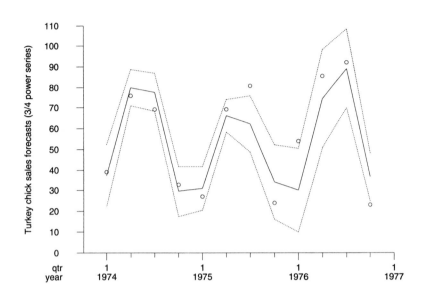

Figure 4.12 Forecast analysis including 1975/3.

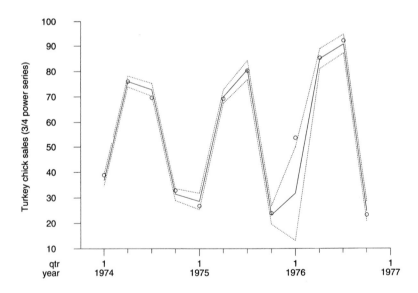

Figure 4.13 Fitted estimates at end 1976.

4.3 Retrospective Perspectives

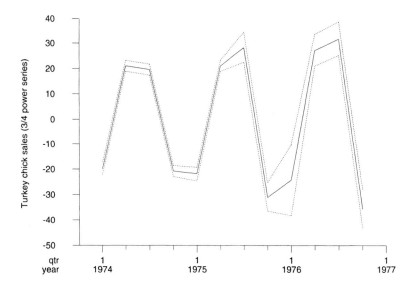

Figure 4.14 Estimated seasonal pattern at end 1976.

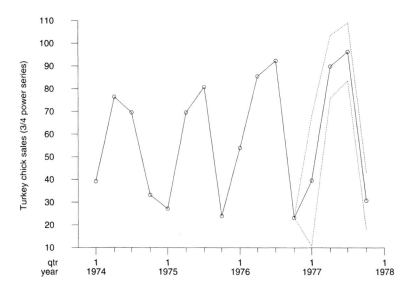

Figure 4.15 Forecasts for 1977.

114   Applied Bayesian Forecasting and Time Series Analysis

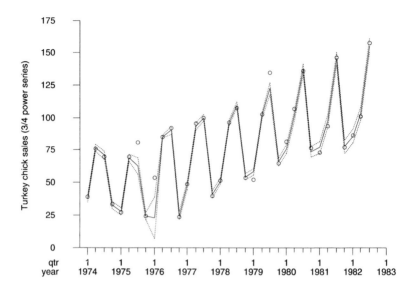

Figure 4.16 Retrospective analysis from 'live' on-line.

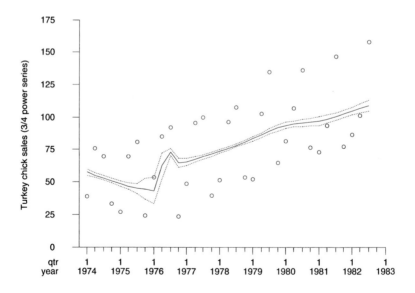

Figure 4.17 Estimated level from 'live' on-line.

4.3 Retrospective Perspectives

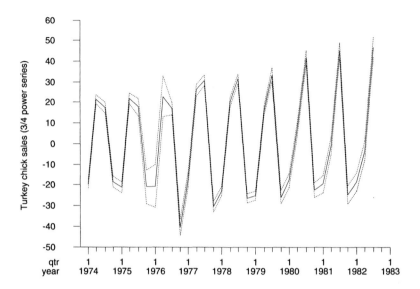

Figure 4.18 Estimated seasonal pattern from 'live' on-line.

Figure 4.16 shows the retrospective analysis corresponding to the live forecast analysis in Section 4.2. As you can see, the correspondence of the estimated values and sales values is extremely good with the exception of the three very high outlying values omitted from the on-line analysis.

Figure 4.17 shows the estimated level. The pattern of change through 1976 does not seem to ring true: sharp increases in the second and third quarters followed by an equally sharp decrease in the fourth quarter. Indeed, it is apparent from our conclusions about the nature of the seasonal pattern in 1976 that these estimated level changes are confounded with the seasonal component estimates. Notice that the estimated seasonal pattern for 1976 in Figure 4.18 maintains the quarter three effect lower than the quarter two effect when we have already concluded that the reverse was in fact true. These observations reinforce the point we made in the previous section: there is scope for improved estimation based upon what we believe having seen the pattern of the data for 1977 and beyond. It seems likely that properly attributing the movement in the series during 1976 to the seasonal component will lead to a much smoother, more realistic estimated trend.

Figures 4.19 through 4.21 show the fitted estimates for sales, and the estimated trend and seasonal components from a retrospective analysis that includes the discoveries made during the foregoing reassessment. The modified on-line analysis follows the live analysis except that the sales values

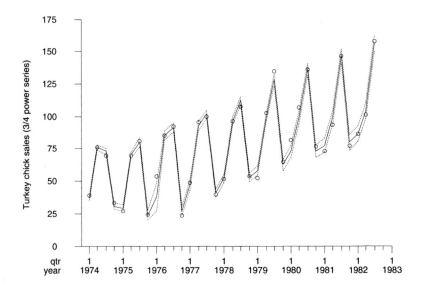

Figure 4.19 Fitted values: the view from 1984.

for 1975/3 and 1979/3 are not omitted. Component priors are made more uncertain at these times to allow adaptation to the system changes.

The fitted estimates, Figure 4.19, for all points except the first quarter of 1976—which is still regarded as a maverick point, the high volume of sales being a fluke for that period and hence omitted from the estimation here—are very good. The estimated level in Figure 4.20 shows a turn around from decline to consistent positive growth of around 16 ($8^{4/3}$) units per year (on the real turkey sales scale) at the time of the first seasonal pattern change in the third quarter of 1975. The changes in seasonal pattern are readily apparent from Figure 4.21, particularly the switch from a square periodic form in the early years to a triangular form in the later years, and a concomitant increase in the peak to trough variation. The alternative view in Figure 4.22 gives a convenient way to see the pattern and scale of the changes. (See the exercises for some interesting remarks about estimated seasonal factors in this graph.)

4.3.3 Final Remarks

It is important to keep in mind the nature of the alternative analyses seen in this section. The live forecast analysis is the best we could do *at the time*. The on-line 'forecast' analysis done at a later time is how we could have

4.3 Retrospective Perspectives

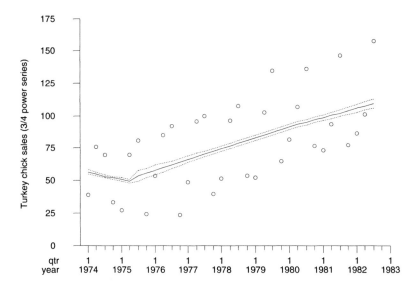

Figure 4.20 Estimated level: the view from 1984.

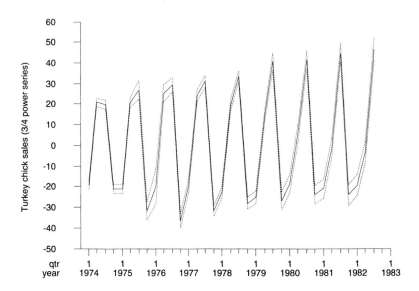

Figure 4.21 Estimated seasonal pattern: the view from 1984.

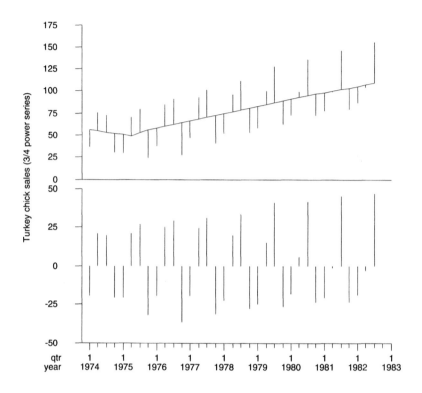

Figure 4.22 Estimated seasonal pattern: the view from 1984.

done in the live situation if we had the benefit of foresight with respect to the nature of the 'unusual' sales values. The retrospective analysis shows how we could do with the benefit of complete foresight, or equivalently, what actually happened.

When taking a view of what actually happened in the past we start from a position of greater knowledge than at any time during the development of that past. Such extra knowledge can be used to clarify the nature of unusual activity in a system, to obtain a better informed idea of the present position, and thereby potentially improve future forecasting.

4.4 Summary

In this chapter we examined the problem of forecasting sales of turkey chicks in Eire. We saw how data transformations can be usefully applied to remove complicating nonlinearities in a series, how to assess the merits of alternative transformations, and how to select a suitable transformation scale on which to analyse the data.

We analysed the turkey series transformed according to a power law with index 3/4, always remembering that forecasts and estimates of change are to be communicated on the original scale for interpretation and assessment. Changes in the underlying structure of the series were identified, analysed, and estimated.

We saw how a live forecasting system can identify unusual points and how a forecaster should and can respond in a dynamic system when they occur. Finally, we examined how the past may be assessed, and reassessed, and how such assessments can be used to improve future forecast performance.

4.5 Exercises

Exercise 4.1 Examine alternative transformations to the square root and 3/4 power transformation. Compare your results to the analyses reported in the text.

Exercise 4.2 Analysis of a series transformed to the square root scale reveals a pattern of approximately constant growth, for example of two units per year. What are the implications of this observation for the raw data series?

Exercise 4.3 Refer to Section 4.1.3. Is a single harmonic model appropriate for either the pre-1979 or post-1979 data segments?

Exercise 4.4 Refer to Section 4.2. Explore alternative discount factor settings in the live analysis. You will need to make your own assessments in response to monitor signals where these occur differently from the analysis in the text. Notice how the forecast monitor reacts to the changes.

Exercise 4.5 Analysis of forecast residuals did not feature explicitly in the text. Simulating the live environment of Section 4.2, stop at each break point and examine the forecast residuals from the analysis to that time. Would you make any changes to the analysis reported? Examine the forecast residuals from the analysis of the full series. What do you think of these? What do you conclude from this study and the results of the previous exercise?

Exercise 4.6 Perform the forecast analysis from 1977 using priors from the retrospective analysis to the end of 1976 discussed in Section 4.3. Comment upon the differences between this analysis and the analysis in Section 4.2 paying particular attention to forecast uncertainties in 1977 and 1978.

Exercise 4.7 Refer to Section 4.3. Explore retrospective estimates for alternative discount factor settings. Comment on the sensitivity of the component estimates, especially estimates of structural change, to different discounts.

Exercise 4.8 Take the fitted estimates from the retrospective view from 1984 and transform to the observation scale. What do you conclude about the rate of growth of turkey chick sales from 1976 to 1983?

Exercise 4.9 Refer to Figures 4.20 and 4.21. Do you feel that toward the end of the series the trend component is underestimated and that the seasonal component estimates are compensating for the effect? Examine the pre- and post-1979 portions of Figure 4.21. What do you think now? Note the form of the seasonal pattern post-1979: three negative factors (quarters 1, 2, and 4) and a single positive factor (quarter 3). Do you still think that the trend is underestimated?

Exercise 4.10 Refer to Figure 4.22. The estimated seasonal factors for 1981 (the last full year in the data set) from the retrospective analysis are -20.8, -1.3, 45.0, and -24.0. Notice that these do not sum exactly to zero. The estimated seasonal factors at the end of 1981 are -19.4, -2.0, 45.4, and -24.0, which do sum to zero. Explain these observations. [**Hint:** Think about the contrast of dynamic and static models.] Would you expect to observe the same effect with a static model? Use BATS to confirm your conclusion.

Exercise 4.11 Turkeys grown for consumption are slaughtered 20 (± 2) weeks from hatching. Traditional turkey consumption festivals like Christmas and Easter naturally lead to increased sales of fresh and frozen turkeys. Can you find any evidence of this effect in the turkey chick sales series?

Exercise 4.12 Research project: try to discover reasons for the observed changes in seasonal pattern in the turkey sales series, and for the unusually high volume of sales in the first quarter of 1976. Make forecasts for 1984. Obtain turkey chick sales figures for 1984 to the present and assess your 1984 forecasts. Analyse the series since 1984 and comment on your results. Send your discoveries and the latest data to the authors.

Chapter 5

Application: Market Share

In this chapter we analyse market share for a consumer product. The product is widely known world wide and the brand leader in its major markets. The study will focus on one of those markets. We build a model for market share that utilises available information on product price and measures of promotional activity. The objectives are to determine a model with good predictive power and to assess the importance of suggested explanatory variables. In addition, we use the example to highlight interesting and important aspects of regression modelling.

5.1 Exploratory Analysis

Figure 5.1 shows four weekly series for 1990 and 1991: percent market share for a consumer product, the price of the product relative to an index of competitors' prices, a measure of the producer's promotional effort, and a measure of competitors' promotional effort.

The product is the clear market leader with market share varying between 39% and 44% over the two years 1990–91. Although not apparent from the figure there are several competitors in the market and the degree of competition is high.

Eyeballing the series one can discern an inverse relationship between market share and relative price, just as economic theory predicts for a competitive market. The relationship between promotional effort and market share, however, is not so easily discernible. This is not a surprising observation

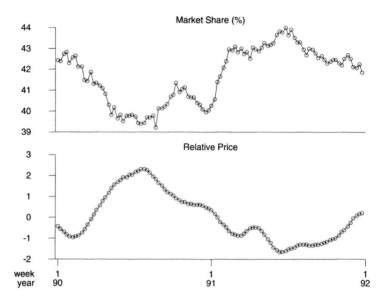

Figure 5.1a Market share and relative price.

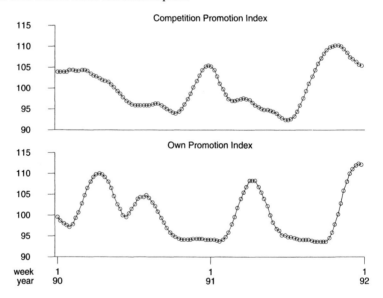

Figure 5.1b Competitors' and own promotion indices.

5.1 Exploratory Analysis

when the nature of the promotion indices is borne in mind. Both indices are constructed according to models currently favoured by the company. Inputs into the models include the number of adverts placed in newspapers and magazines (weighted by circulation) and on television, and the number of days on which adverts appeared during the week (there are cumulative exposure and memory effects in the models) among others.

Unlike price, consumers do not see the indices; they see the raw advertising and respond in myriad complex ways psychologists are actively trying to understand. Ideally we would have a model that takes the actual promotional details as input and yields effect on consumer demand as output. In the absence of such detailed information we work with the raw promotion activity indices as a proxy.

We remarked that relationships between market share and the promotion indices are not easy to see on visual inspection. One of the objectives of the analysis in this chapter is to determine and assess these relationships. The first question to answer is whether there are any relationships at all. If there are, how strong are they? How stable? How do they compare and contrast?

Normal expectations are that market share will increase in response to product advertising, and conversely that market share will decrease in response to competitor product advertising. Several alternative models may be posited to represent these relationships. The simplest, and most obvious, supposes additive effects for each index,

$$\text{market share} = \text{level} + \alpha \text{PROM} + \beta \text{CPROM} + \text{other components},$$

where PROM is the product promotion index and CPROM is the competition promotion index. A special case of interest here is the case where only the difference between product and competitor promotional effort is significant: $\alpha = -\beta$.

An alternative model is to suppose that market share varies with the reciprocal of competitors' promotional activity,

$$\text{market share} = \text{level} + \alpha \text{PROM} + \gamma \text{CPROM}^{-1} + \text{other components}.$$

An interesting special case is where only relative promotional effort is relevant: $\alpha = \gamma$.

In this chapter we study the first of these models, where the promotional effort indices are additive explanatory variables. Exploration of alternatives is pursued in the exercises.

5.2 Dynamic Regression Model

The model we investigate in this study is a two component model in which the components are a level and a three variable regression,

$$\text{market share}_t = \text{level}_t + \alpha_t \text{PRICE}_t + \beta_t \text{PROM}_t + \gamma_t \text{CPROM}_t,$$

where PRICE is measured relative to a measure of competitors' average prices; PROM and CPROM are producer and competitor promotion indices as already remarked. The level and regression coefficients have a steady (random walk) evolution; that is, in normal circumstances the coefficients are expected to be broadly similar from week to week. In terms of the standard DLM notation the model is represented by the observation and system equations

$$Y_t = F_t \theta_t + \nu_t, \qquad \nu_t \sim N[0, V_t],$$
$$\theta_t = G_t \theta_{t-1} + \omega_t, \qquad \omega_t \sim N[0, W_t].$$

where the response, regression vector, and system matrix are identified as follows:

$$Y_t = \text{marketshare}_t,$$

$$F_t = \begin{pmatrix} 1 \\ \text{PRICE}_t \\ \text{PROM}_t \\ \text{CPROM}_t \end{pmatrix}, \quad G_t = \begin{pmatrix} 1 & 0 & 0 & 0 \\ 0 & 1 & 0 & 0 \\ 0 & 0 & 1 & 0 \\ 0 & 0 & 0 & 1 \end{pmatrix}, \quad \theta_t = \begin{pmatrix} \text{level}_t \\ \alpha_t \\ \beta_t \\ \gamma_t \end{pmatrix}.$$

The observation error variance, V_t, will be estimated along with the other model parameters; the evolution innovation variance, W_t, is specified using the block discounting technique. Since there are two component blocks in the model, the trend and the three variable regression, the evolution variance matrix is a block diagonal matrix with two blocks,

$$W_t = \begin{pmatrix} \text{trend block} & 0 \\ 0 & \text{regression block} \end{pmatrix}.$$

The component blocks are determined by the component discount factors applied to the previous week's posterior covariance matrix. With the posterior covariance matrix partitioned into trend and regression component blocks,

$$C_t = \begin{pmatrix} C_{T,t-1} & C_{TR,t-1} \\ C_{RT,t-1} & C_{R,t-1} \end{pmatrix},$$

the evolution covariance matrix is

$$W_t = \begin{pmatrix} (\delta_T^{-1} - 1)C_{T,t-1} & 0 \\ 0 & (\delta_R^{-1} - 1)C_{R,t-1} \end{pmatrix},$$

where δ_T is the trend component discount factor and δ_R is the regression component discount factor.

Table 5.1 Initial prior setting

Component	Mean	Std. Dev.
Level	42	5
Price	0	2
Promotion Index	0	2
Competition Promotion Index	0	2

Observation std. dev. estimate 1, degrees of freedom 1

5.3 A First Analysis

We begin with a static analysis of the level plus regression model, a bench mark against which to assess subsequent analyses. The initial priors used in this analysis are given in Table 5.1, having been specified to represent considerable prior uncertainty. An approximate 95% interval for the level, mean plus or minus two standard deviations, of [32, 52] encompasses the entire range spanned by the data. That is surely an uncertain prior estimate. Do you agree with our claim that the prior settings for the other model components represent great uncertainty?

Figure 5.2 shows the one week ahead forecasts from the static analysis. The 90% forecast uncertainty limits look to be a little on the large side. A more troubling observation is that the point forecasts are undeniably missing some of the series structure. The forecasts are consistently high in the latter part of 1990 and consistently low throughout 1991 (with one or two exceptions). The missing structure is confirmed in Figure 5.3 which indicates a strongly nonrandom autocorrelation pattern in the forecast residuals. Positive autocorrelation in forecast residuals is typically an indication of insufficient dynamic movement in component parameters, that is, discount factors that are too high. In this analysis there is no dynamic movement at all, so it is not surprising to find evidence of too high discounts!

Another troubling result is the on-line estimated level, shown in Figure 5.4. The estimated value develops in what is surely a ludicrous manner. What is more, the uncertainty around the estimates is huge even when all 104 observations have been processed. In the discussion of the forecast residuals we concluded that the discount factors are too high; now we observe too much uncertainty in (at least) one component estimate. What can this apparently conflicting evidence mean? Very obviously there are some modelling problems to resolve.

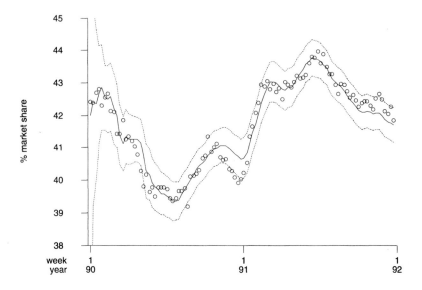

Figure 5.2 Forecasts for market share.

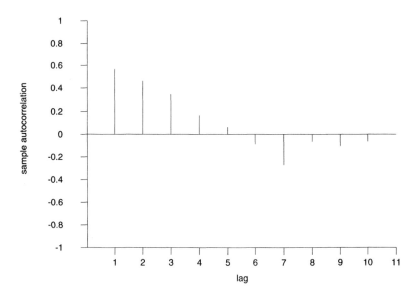

Figure 5.3 Autocorrelation of forecast residuals.

5.3 A First Analysis

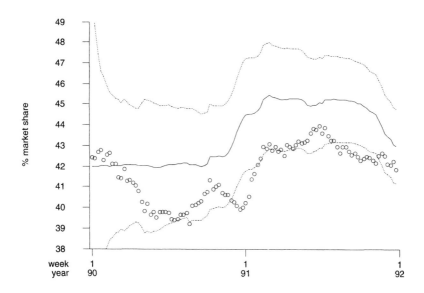

Figure 5.4 On-line estimate of underlying level.

5.3.1 Scaling Regression Variables

Consider the problem with the level estimate first. Recall that the posited model says that market share is comprised of a level and contributions from the relative price of the product and measures of promotional effort for the product and for competing products. We need to think a little more carefully about what we mean by this model. What do we interpret by the level of market share? And what do we mean by the contributions of price and promotion?

The natural interpretation of the level is an underlying value in the absence of other effects: the value of market share we expect to see in the absence of contributions from price and promotion. The difficulty we have observed with the level estimate here arises from the fact that the promotion variables are measured on an arbitrary scale. There is always some level of promotion (in this data set) and so estimation of the level component is confounded with the underlying level of promotion. It is simply not possible to estimate the absolute effect of promotion with such data. The best we can hope to do is to estimate an underlying level of market share corresponding to the underlying level of promotion, together with the contribution to market share of changes from that underlying level of promotion.

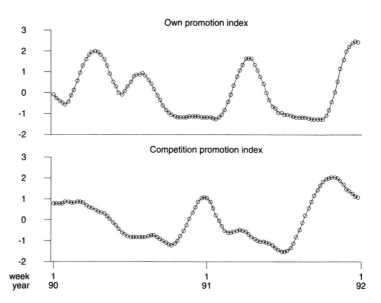

Figure 5.5 Standardised promotion indices.

Centreing the promotion variables is the way to proceed. With a zero mean value the average contribution is nothing, which means that the level component is now the only source of underlying market share remaining in the model.

The importance of communication must not be underplayed in this matter. We are concerned here with interpretation of 'the level' and have shown that it is a level *dependent on the other variables in the model*. Thus its meaning is always relative to a given price/promotion (in this case) setting and this must always be made clear to management. Generally we want the level to be relative to 'normal' operating conditions, which usually corresponds to average values for inputs. Centreing achieves precisely that requirement.

Formally the model structure is identical to the structure set out above. All that has changed is the scale on which the promotion variables are measured. Figure 5.5 shows the rescaled promotion measures. Compare these with the original scaling shown in Figure 5.1.

It may seem at this point that we arbitrarily picked on the promotion indices on which to lay blame for the poor level estimation. Does the price variable not bear any responsibility? Well, in fact the relative price measure is already centred as you can see from Figure 5.1. Incidentally, it is often useful to standardise explanatory variables to have unit variance. Standardisation puts regressors on an equal footing, facilitating comparison of coefficients. Figure 5.5 actually shows standardised promotion indices.

This section has taken a rather indirect approach to pointing out the problems with uncentred regression variables in models having separate trend components. We could have given the latter discussion before attempting any analysis, then immediately produced the nice results of the next section. In practice one always subjects tentative models to close scrutiny before attempting analysis. However, experience has taught us that the pedagogical device of showing obviously poor results obtained from seemingly satisfactory modelling is extremely effective. Lessons starkly illustrated are not soon forgotten. Moreover, in complex models, problems such as level confounding may not be apparent even on close scrutiny before analysis is undertaken. It is important to be aware of the possible consequences of model misinterpretation so that when problems are observed in results, one does not have to search blindly for potential causes.

5.4 Analysis with Rescaled Promotions

A static analysis of the market share model with centred promotions is a good place to begin. However, we have already seen a similar analysis for the uncentred promotions model in the previous section. Having addressed the issue of poor level estimation evident in that analysis, the problem of serially correlated forecast residuals remains. We leave it as an exercise for the reader to confirm that regressor centreing obviates the level estimation problem.

If the model form is basically correct, serial correlation in forecast residuals is typically the result of insufficient dynamic movement in component parameters. If the model form is deficient, the serial correlation will remain a feature of the dynamic analysis. Let us determine which is the case for the market share model.

5.4.1 Dynamic Analysis

It is often the case with dynamic models that regressions on explanatory variables exhibit less temporal movement than do pure time series components such as trends. This leads us to consider first models with regression component discounts greater than trend component discounts. In the present case, however, investigation of a range of alternative sets of component discount factors reveals that the regression component is actually more unstable in its contribution to the model than the underlying trend.

Table 5.2 Forecast performance summary[†]

Regression Component Discount	Trend Component Discount			
	0.95	0.98	0.99	1.00
0.90	0.0611 0.1894 −22.0000	0.0592* 0.1865 −13.3400	0.0593 0.1874 −12.1800	0.0601 0.1889 −11.9000*
0.95	0.0610 0.1863* −12.0800	0.0633 0.1875 −12.2500	0.0646 0.1895 −12.9900	0.0664 0.1932 −14.0700
0.98	0.0691 0.2009 −16.8900	0.0757 0.2120 −20.9800	0.0786 0.2173 −22.5900	0.0820 0.2247 −24.3200
1.00	0.0817 0.2240 −25.4200	0.0948 0.2429 −32.1700	0.1013 0.2554 −35.1400	0.1108 0.2680 −39.1900

[†] mean squared error (mse), mean absolute deviation (mad), and log predictive likelihood.
* Best value for measure.
Observation variance discount is 0.99.

Table 5.2 shows forecast performance summary statistics for several sets of discount factors. As you can see, the best performing analyses are those wherein the regression component discount factor is around 0.9 and the trend component discount is close to 1.

Table 5.3 shows another set of forecast summary statistics. This time the analyses were performed with the BATS forecast monitor on. Several of the analyses were subject to monitor warning signals (not always at the same times) and in each case the program default action was selected. (The BATS forecast monitoring scheme is detailed in Chapter 11.) As you can see, intervention analysis is typically superior to the nonintervention analysis whatever the discount factors. The best analyses result from similar combinations of discount factors in both intervention and nonintervention analysis.

One possible reason why the regression component seems to be more prone to change than is the level is the artificial nature of the promotion measures. These constructed variables are the result of some considerable modelling assumptions and approximations as we have already remarked. Price is not constructed in the same way but it is quite possible that the 'true' price-demand relationship is only approximated by the simple regression

5.4 Analysis with Rescaled Promotions

Table 5.3 Forecast performance summary with intervention[†]

Regression Component Discount	Trend Component Discount			
	0.95	0.98	0.99	1.00
0.90	0.0611 0.1894 −22.0000	0.0544 0.1801* −10.7600	0.0543* 0.1810 −8.9900	0.0545 0.1823 −7.8200*
0.95	0.0610 0.1863 −12.0800	0.0573 0.1812 −9.1300	0.0583 0.1822 −9.2400	0.0598 0.1845 −9.8600
0.98	0.0672 0.1979 −16.2900	0.0662 0.1927 −13.9900	0.0673 0.1934 −15.7600	0.0693 0.1950 −16.6500
1.00	0.0778 0.2175 −23.0700	0.0738 0.2043 −19.8100	0.0757 0.2088 −20.1500	0.0824 0.2175 −24.3500

[†] mean squared error (mse), mean absolute deviation (mad), and log predictive likelihood.
* Best value for measure.
Observation variance discount is 0.99.

we have used, particularly where there are substantial changes in price. Further investigation of this point is possible by generalising the model to multiple regression components where the price component discount factor may be varied independently of the promotion component discount factor, an idea we expand upon below.

Figure 5.9 shows the estimated regression parameters. The promotion measure coefficients do not show much evidence of instability; the evidence is that there is somewhat more movement in the price variable coefficient.

Figures 5.6 through 5.9 show selected results from the intervention analysis with discount factors 0.98 for the trend, 0.95 for the regression, and 0.99 for the observation variance. The one step forecast development is shown in Figure 5.6. The point forecasts look very reasonable; the 90% uncertainty limits are very wide to begin with—reflecting the great uncertainty in the prior specification—but collapse down quickly as information is sequentially processed. Notice the massively increased uncertainty at week 35 of 1990 following the omission of the outlier at the preceeding week and the automatic program response invoked following the forecast monitor signal. (The monitor response mechanism is described in Chapter 11. Briefly, the automatic action in response to an outlier is to substantially increase prior

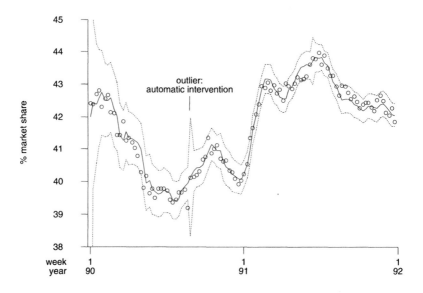

Figure 5.6 Forecasts from intervention analysis.

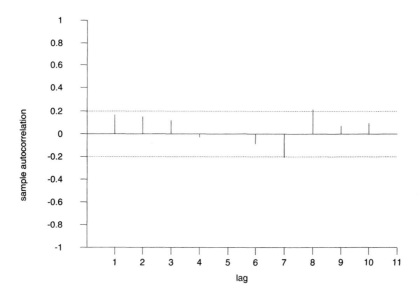

Figure 5.7 Autocorrelation of forecast residuals.

5.4 Analysis with Rescaled Promotions

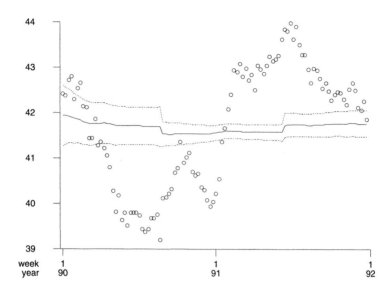

Figure 5.8 Estimated underlying level.

uncertainty on every model component for the next week. That increased uncertainty is directly reflected in the resulting forecast.) When the observation for 90/35 is processed, component uncertainties are greatly reduced (almost back to the preintervention level) as the observation is in good agreement with the forecast. A second monitor signal was issued at 91/23 and you can see that the forecast uncertainty at the following week is increased. On this occasion, however, the increase is of a smaller magnitude than before. Here, the signal is of a different kind. There is not a single large difference between the forecast and the outcome at 91/23 but cumulative evidence of poor forecasting over a couple of weeks. Therefore, the observed value at 91/23 is not ignored (as was 90/34) and the information it provides reduces the posterior uncertainty at that time. The prior uncertainty for 91/24 is proportionally increased over the posterior at 91/23 by the same factor as the prior uncertainty at 90/35 is increased over the posterior at 90/34. But since the posterior uncertainty at 91/23 is smaller than that at 90/34, the resulting forecast uncertainty at 91/24 is smaller than it is at 90/35.

The sample autocorrelations of the forecast residuals are shown in Figure 5.7. They reveal no obvious evidence of systematic over or under forecasting in contrast to the original analysis. The autocorrelation estimates at lags seven and eight do exceed the approximate 90% limits, but only by a tiny amount. Certainly not by enough, alone, to prompt further consideration.

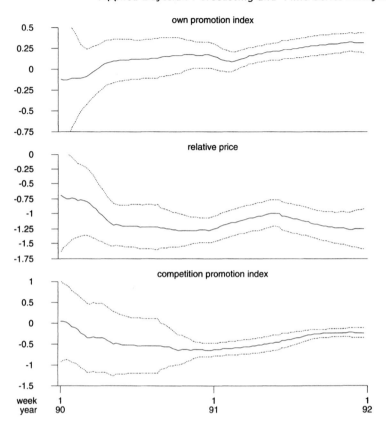

Figure 5.9 Estimated regression coefficients.

The estimated level in Figure 5.8 is extremely stable, as indeed it must be with a high discount factor. Where substantial change is possible—at the times of automatic intervention when the prior uncertainty is substantially increased—there is an interesting effect. A (small) downward shift in the level occurs at 90/35, offset by a corresponding upward shift at 91/24. In absolute terms the drop is very small, from 41.7% to 41.5%. The question is, do we believe the effect? The estimated uncertainty in the level is quite consistent with the hypothesis of an unchanged level. If there were other market factors active over the period, such as a short-lived new entrant, or production problems for example, then the effect might be considered real. In this case, however, there are no such external factors (that we know about) and it seems most likely that the observed effect is an artefact of the intervention analysis.

Figure 5.9 shows the estimated regression coefficients. Notice first that there is no sudden change in any of the coefficients at the intervention points. Why not? There are changes in the level and we just dismissed them as not real, the product of the intervention. Should we not observe something similar with the regression component? The answer lies in the mode of the interventions. We let the program perform intervention according to the default rules built in. Default intervention component discounts are 0.1 for trends and 0.8 for regressions reflecting the typical pattern of more robustness in regressions than in trends. This means that whatever movement there is in the data may more easily feed into the trend than into the regression. Since the movement is quite small and not obviously a regression change, that is precisely what happens. Had we been more informative with the interventions—recognising the stability in the level—there would be nothing to discuss here as we will see in the next section.

All three estimated regression coefficients exhibit systematic evolution. The price variable, as expected, is a significant component of the model, negatively related to market share. Notice that the coefficient is greater than 1 in magnitude (ignoring the first few weeks in 1990). This is indicative of a highly elastic demand, typical of a competitive market. A 1% increase in relative price results in a greater percentage reduction in market share. (See also Exercise 5.8.)

There is a good deal of uncertainty surrounding the promotion coefficients in the early part of the series. Company promotion efforts may reasonably be judged to have no effect in 1990, and competitors' promotions to have no effect until September 1990. We will say more about the regressions in the next section where the previously detailed shortcomings of the present analysis with respect to the level and interventions are addressed.

5.5 Final Analysis

In this section we address the deficiencies of the analysis of the previous section regarding treatment of the trend component and interventions. The point identified as an outlier, week 34 of 1990, is excluded from the analysis without increasing prior uncertainties for 90/35. No intervention is performed at 91/24. The forecast monitor does not have any foresight: a signal is an indication of *potential* change only. After examining the circumstances leading to the signal it was decided that the recent observations were noisier than the monitor sensitivity allowed but nothing fundamental had changed. We illustrate the analysis using component discount factor 1.0 for the trend, 0.9 for the regression, and 0.99 for the observation variance. It is a useful and instructive exercise to examine results for alternative discount factors; we leave that to you.

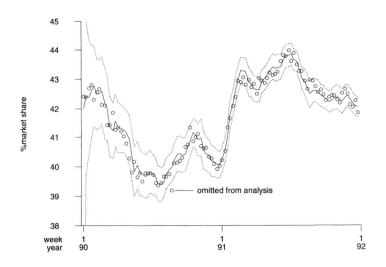

Figure 5.10 Market share forecasts from final intervention analysis.

One step ahead forecasts are shown in Figure 5.10. Comparison with the forecasts from the previous analysis in Figure 5.6 reveals only very minor differences with the exception that the current analysis does not exhibit the great forecast uncertainties at 90/35 and 91/24. This is also revealed in the summary statistics which are mean square error = 0.056, mean absolute deviation = 0.185, and predictive log-likelihood = -7.5. The measures of point forecast performance are almost unchanged but the predictive log-likelihood is improved by the much less diffuse forecast distributions at 90/35 and 91/24. Notice in Figure 5.10 that, apart from the outlier, no points lie outside the 90% prediction intervals—we would expect about 10 to do so. This is an indication that the combination of discount factors can be improved. See also the comments about the effectiveness of the regression components and their contribution to forecast variance.

The estimated level in Figure 5.11 is constant (because the discount factor of unity defines the component to be static) at 41.68 with 90% probability interval [41.51, 41.85].

The forecast residuals exhibit no autocorrelation, the sample values being almost identical to those of the previous analysis shown in Figure 5.7. A normal quantile plot of the standardised forecast residuals, Figure 5.12, shows no evidence of non-normality.

5.5 Final Analysis

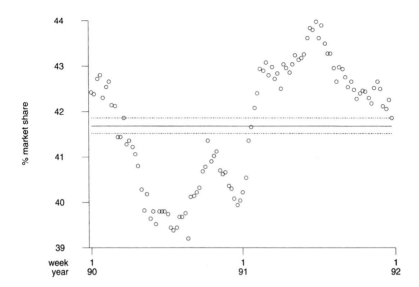

Figure 5.11 Estimated underlying level from final intervention analysis.

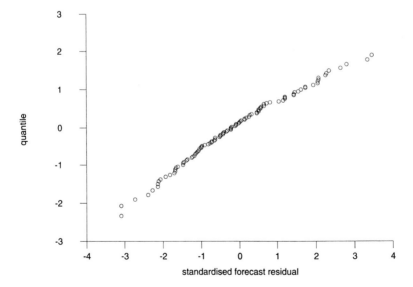

Figure 5.12 Normal quantile plot of standardised forecast residuals.

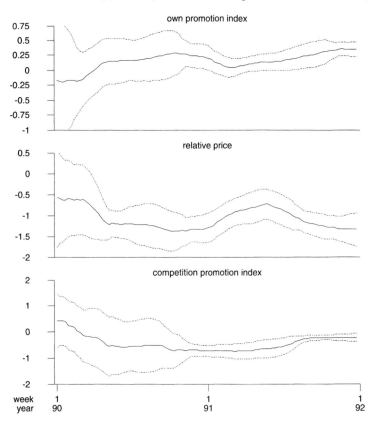

Figure 5.13 Estimated regression coefficients.

Finally, a few comments on the estimated regression coefficients in Figure 5.13. The promotion coefficient varies between 0 and 0.25; the 90% uncertainty bounds are consistent with the coefficient being zero for almost the entire two year period. This suggests that the company's promotional activities have little effect on market share. (Actually the conclusion is not quite so strong. Recall the discussion on level confounding and regressors. What we should have said is that variations in promotional effort around the 'average' level have no impact. It is not at all clear from the available data what would happen to market share if the company ceased all promotional activity.)

The coefficient on competitors promotional activity is really not distinguishable from zero with any confidence until the last few weeks of 1990. Thereafter it is stable around -0.5 until the middle of 1991 after which it slowly moves to -0.25 where it stays for the final third of that year.

5.6 'What if?' Projections

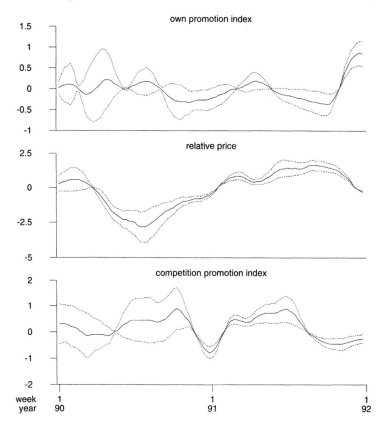

Figure 5.14 Estimated regression effects.

Interestingly the diminishing impact of competitor promotion as measured by the regression coincides with a sharp rise in promotional activity.

The contribution of the price variable to the model is substantial, the estimated coefficient varying between -1.4 and -0.8. The coefficient decreases (in magnitude) during the first half of 1991, then increases back to its 1990 level. This change largely coincides with a period of increased company promotional activity. It also coincides with a period of relative price advantage. Figure 5.14 shows the effects of the regression variables (that is, the variable multiplied by its estimated regression coefficient).

5.6 'What if?' Projections

Forecasting with models that include regression components is more demanding than with nonregression models. It is necessary to know in advance the values that the regressor variables will assume in the forecast periods. Where those values are known or can be controlled there is no difficulty. However, in many cases, the present case included, the values are not known and cannot all be determined in advance by the forecaster.

The product's price is under control of the company. Competitors' prices, however are not. Since those prices are a component of the relative price index used in the forecast model, there is a problem. We can only guess at what the prices will be. Similarly, the company's future promotional effort is controllable but that of competitors is not. Thus, it is not feasible to make predictions based on the constructed model without first making predictions about the values of some of the model inputs.

In practice one may have other models that may be used to forecast the inputs required. However, since those forecasts are uncertain, using point predictions as inputs into the market share model will give erroneously confident forecasts. It is vital to incorporate the uncertainty about the future values of the predictors into forecasts constructed using those predictors.

It is, of course, entirely possible to construct a nonlinear model in which the unknown predictors are modelled with probability (forecast) distributions, but that is a complicated undertaking and outside the scope of this book.

There is a much simpler alternative to nonlinear modelling that remains entirely within the linear world. For any given set of predictor values, or series of sets of values when forecasting several periods ahead, the model routinely produces forecasts. We do not know which set of values the predictors will manifest, but often we can be reasonably certain about ranges of values within which the actual values will lie. By taking sets of values representing these ranges and producing forecasts from each, we construct a range of forecasts. The envelope of these 'What if?' forecasts provides an overall forecast in which all uncertainties are considered: uncertainties about model parameters and uncertainties about future values of predictor variables.

The envelope approach is only a rough approximation since no account is taken of the probabilities of the alternative scenarios. However, it is an extremely useful and practicable technique. It has the considerable advantages of being readily implemented and understood. Moreover, contrasts among the individual scenario forecasts often provide valuable insight.

'What if?' forecasts are also used for assessing the impact of proposed courses of action. In that context it is not the envelope of the resulting forecasts that is of interest, rather it is the comparison of the individual forecast scenarios. We have already seen this mechanism in action in Chapter 4 in the context of assessing the implications of possible interventions.

5.7 Contemporaneous and Lagged Relationships

Table 5.4 Forecasts for 1992

Week	Mean	90% Margin
	Scenario 1	
1992/1	41.40	0.359
1992/2	41.40	0.362
1992/3	41.40	0.366
1992/4	41.40	0.369
1992/5	41.40	0.372
	Scenario 2	
1992/1	41.36	0.365
1992/2	41.31	0.379
1992/3	41.25	0.395
1992/4	41.23	0.404
1992/5	41.19	0.417
	Scenario 3	
1992/1	41.23	0.390
1992/2	41.22	0.396
1992/3	41.22	0.401
1992/4	41.22	0.407
1992/5	41.20	0.415
	Scenario 4	
1992/1	41.19	0.399
1992/2	41.13	0.419
1992/3	41.07	0.439
1992/4	41.05	0.452
1992/5	41.00	0.474

Table 5.4 shows the week-by-week forecasts for the first five weeks of 1992 for four different promotion scenarios, relative price in all cases being fixed at 0.206, the final value for 1991:

- company and competitor promotion indices set to 0;
- company promotion index PROM set to its first five values of 1990, competitor promotion index CPROM set to 0;
- company promotion index PROM set to 0, competitor promotion index CPROM set to its first five values of 1990;
- both promotion indices set to their first five values of 1990.

We make no further comment on these results. Instead we encourage you to study the conditions upon which each set is predicated and consider the implications of each as shown in the table.

5.7 Contemporaneous and Lagged Relationships

In the analysis we have presented for the market share data there is no consideration of the possible explanatory power of time lagged variables. There are three reasons for the omission.

First, the nature of the data itself: market share for a consumer product. By definition consumer products like foodstuffs are purchased regularly and often. Purchasing decisions are made on the basis of current price: what do I care how much a loaf of bread cost yesterday (or what it might cost tomorrow) if I need a loaf today? The same argument does not hold true for nonconsumable products. Potential purchasers of nonconsumables may be interested in previous prices as an indicator of possible future price trends. Items that do not have to be purchased today can be purchased tomorrow or next month if price is expected to fall.

Second, the timing period of the data: one week. While there is undoubtedly a memory effect day-to-day from advertising exposure it is very unlikely that such an effect will be noticeable over a week or more.

Third, authors' expediency: we want to leave something for you to do in the exercises!

5.8 Multiple Regression Components

The model analysed in this chapter defines a single regression component comprising three variables: price and two promotion indices. An alternative to the single regression component model is a model with multiple regression components. The difference between single and multiple regression components is similar to the difference between components of contrasting type: trend and regression, or trend and seasonal, for example. Specifically, in the case of a single regression component all of the regressor variables are treated as a group. When computing evolution variance a single discount factor is applied to the regression *block* covariance matrix preserving correlation structure between the regressors. Correlations between components, on the other hand, are reduced through the block discounting evolution scheme. See Sections 3.6 and 5.2 for details.

The implication of the block discounting scheme for multiple regression components is that different kinds of regressors may be considered separately. For the market share series, price may be separated from the promotion indices. In this way it is possible to allow the price component to be more or less stable than the promotion indices component, a feature that is suggested from the analysis above. Moreover, separating price from the indices decouples the two types of variable (resets correlation to zero) at times of intervention. This can be important if there are changes in the type of advertising for example. Increased uncertainty from a new kind of advertising strategy can be prevented from inadvertently, and undesirably, increasing uncertainty on the price-market share relationship.

These ideas are pursued further in the exercises.

5.9 Summary

In this chapter we illustrated modelling and forecasting with dynamic regression component models. We analysed market share of a competitive product in terms of an underlying level, the effect of price relative to competitor prices, and the effects of promotion effort both by the company for the product and by the competition for competing products. The analysis demonstrated the importance of centreing regression variables and the consequences of not doing so. Features and relative merits of single and multiple regression component models were discussed. Finally, we examined the issues of component contributions to forecast uncertainty and the problems inherent in forecasting when regressor values are unknown.

Before we leave the chapter we want to encourage you to read and think about the exercises that follow. Several issues are raised there that we have only briefly touched on in the main text. Nontrivial issues that are important and relevant more widely than to just this particular application. Do examine the exercises...and preferably try some!

5.10 Exercises

Exercise 5.1 Percent market share varies between zero and 100. What are the implications of using the normal distribution for disturbance terms in a model for market share? Is the analysis in the text invalidated by this consideration?

Exercise 5.2 Refer to Section 5.3. The initial static analysis showed strong evidence of serial correlation structure in the forecast residuals and poor level component estimation. In the text the level estimation was addressed first; subsequently the serial correlation was addressed in a revised model. Explore results for dynamic analysis of the initial noncentred regressors model. Comment on component parameter estimates.

Exercise 5.3 *Continuation.* Examine a static analysis of the centred regressors model. Confirm that the estimation problems encountered with the noncentred regressors model do not occur. Confirm also that there is strong residual serial correlation.

Exercise 5.4 Refer to Section 5.4. The forecast summary statistics in Table 5.3 are derived from intervention analysis where the interventions consist only of the BATS automatic adjustment to monitor signals. We stated in the text that different signals occurred depending upon the component discount factors. Explore some of the discount factor combinations in the table and note the times and type of monitor signals generated. Comment on the differences. Do you think there are potential outlier points in the data other than week 34 in 1990?

Exercise 5.5 *Continuation.* The initial priors, Table 5.1, represent very vague beliefs, the level reasonably expected to be anywhere in the interval $[32, 52]$. Experience from 1989 (and earlier) actually gives more guidance, suggesting a level in the range $[40, 44]$. Formulate a prior to represent this information. Using other prior settings and discount factors as in the text, make a comparative analysis. Note where monitor signals occur and the kind of potential breakdown they suggest. Comment on your findings.

Exercise 5.6 When the company's marketing manager reads your technical report and sees the lag eight autocorrelation in the forecast residuals he becomes very excited. He seizes upon the effect you have described as spurious as a missed opportunity. After a few seconds thought he reels off an apparently self-convincing explanation of why there is a real effect of market share two months ago on current market share.

The marketing manager insists that you estimate the effect of including a lagged market share component and produce new forecasts. When he sees your analysis of the model including a regression on centred lagged market share he is suspicious. He demands to know why you did not use the raw lagged value in your modelling. Explain your reasoning.

Exercise 5.7 It is suggested that in addition to price only the *difference* between the producer's promotional effort and that of competitors is significant in determining market share. How would you respond to this suggestion?

Exercise 5.8 Refer to Section 5.4. If the measurement scale of the relative price variable is doubled it is clear that the coefficient estimate will be halved, giving a value less than one. Does that mean we would state that the market is price inelastic? What condition is necessary for the elasticity statement to be correct?

Exercise 5.9 An alternative way of modelling promotional effort is the ratio of the producer's effort to competitor's effort, in other words relative effort. Investigate such a model. What are your conclusions?

Exercise 5.10 Refer to Section 5.7. Studies have shown that there is a memory effect from exposure to advertising. Advertising has an initial

5.10 Exercises

impact and also a decaying impact for a few days thereafter. Repeated exposure to advertising for the same product builds up a stock of advertising awareness (up to some threshold). When advertising ceases, the awareness decays.

The memory effect of advertising suggests that there might exist a relationship between market share and lagged promotional effort. In the text we argue that there probably is not such a relationship in the weekly market share data. Investigate our claim. What are your conclusions?

Exercise 5.11 *Continuation.* Consider the problem of making a forecast of market share for just next week. A model that has only lagged regression variables removes one of the sources of uncertainty from the forecast, namely the uncertainty about the values of those regression variables. What do you think of that argument?

Exercise 5.12 The analyses in the text assume a constant discount factor except at interventions. In some situations it is sensible to change discount factors for several periods at a time. For instance, if a new (or different) kind of advertising campaign is introduced, market reaction could be unusually turbulent for a while. Lowering the discount factor on the promotion component prevents the additional noise in the observation series from incorrectly inflating the observation disturbance term. The reverse effect, periods of increased stability, is also a possibility.

Do you notice any periods of particular stability or variability in the component estimates in the text? Isolating periods of greater and lesser stability, experiment with changing the component discount factors.

Exercise 5.13 Refer to Section 5.6. The forecast uncertainties in each of the scenarios summarised in Table 5.4 increase with each week. Why? Under what circumstances would those uncertainties decrease with lead time? Use BATS to confirm your answers. [**Hint** Think about component discount factors and regressor values and their effects on forecast variance.]

Exercise 5.14 Refer to Section 5.8. BATS does not currently support multiple regression components. Obtain the source code from the authors and implement this extension (or write your own program). Analyse the market share series and determine if separating price from the promotion indices improves forecast performance. Comment on the relative stability of the trend, price, and index components.

Exercise 5.15 Consider the forecast uncertainty after the outlier and automatic intervention for week 90/34 from the following two models:

(i) static trend with automatic exception discount factor 0.1; regression discount factor 0.9 with automatic exception value 0.1;
(ii) static trend and regression with exception discount factors both 0.1.

Explain why the forecast variance after the intervention is greater in the static model than it is in the dynamic regression model. [**Hint** Examine the posterior uncertainties in each model at week 90/34. Examine model performance over the first 34 weeks of 1990.]

Chapter 6

Application: Marriages in Greece

In this chapter we direct the spotlight away from forecasting and focus attention on time series analysis. The objectives of a time series analysis are to identify and estimate patterns of behaviour, to characterise the historical development of time series, and to compare and contrast related series.

The present study examines the annual number of marriages in Greece over the 22 year period 1968 to 1990. Data is available by geographical region (official Greek publications). We examine the regional breakdown later in the chapter. To begin the study we take a look at the aggregate figures for Greece as a whole, shown in Figure 6.1. Leap years are indicated. Several prominent features of the data are easily identified: a strong four year cyclical pattern with a low coinciding with leap years, and a dramatic change in the underlying trend pattern after 1979. Peering a little more closely there is a suggestion that 1972 and 1988 are unusually low even for leap years. Also, the period 1980–1983/4 has the appearance of a transitory phase.

The Greek population expanded from 8.5 million in 1968 to 10 million in 1990, so the decline in marriages cannot be the result of declining population. A more likely cause is cost. Weddings in Greece are expensive occasions (civil weddings became possible only after 1981 and still account for just a tiny proportion of the total), and the Greek economy experienced a serious deterioration during the 1980s.

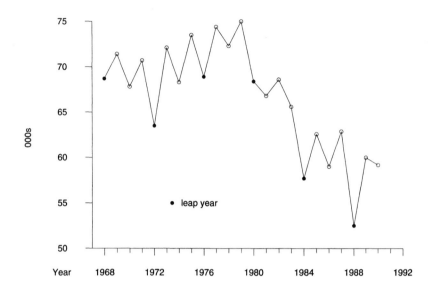

Figure 6.1 Number of marriages (000s) in Greece.

6.1 Analysis I

The preceding discussion suggests that a suitable model for the marriage series will include a linear trend component and a four year cycle component. This is one instance of long term cyclical behaviour having nothing to do with physical systems; it is purely a human psychological phenomenon.

We use a very uncertain initialisation for the analysis: the level mean is set at 70 with standard deviation 1.5, the growth mean is set at 0 with standard deviation 0.5, the cycle factor means are set at 0 with standard deviation 2, and the observation variance is set at 1.5 on a single degree of freedom. This prior specification is for a no-growth trend and gives no information on the cycle pattern (other than overall estimated peak to trough variation implied by the factor standard deviations). Component discount factors are set at 0.95 for the trend and 0.9 for the cycle. The unusually low number of marriages recorded for 1972 is omitted from the analysis. Intervention on the trend component is performed at 1980: the level standard deviation is increased to 5 (the point estimate is untouched at 73), and the growth estimate is reset to -2 with standard deviation 1. This intervention represents a negative outlook tempered by a large degree of uncertainty.

6.1 Analysis I

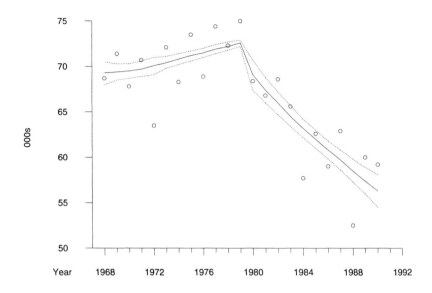

Figure 6.2 Estimated underlying level from analysis I.

Estimated components from this analysis are shown in Figure 6.2 (trend) and Figure 6.3 (cycle), and overall fitted values in Figure 6.4. An alternative view of the estimated cycle component is shown in Figure 6.5.

The overall fitted values exhibit reasonable conformance with the data, the intervention allowing the trend change to be well captured. The estimated trend is well behaved, giving a seemingly excellent summary of the long term series development. The pace of decline during the decade of the 1980s is much greater than the rate of growth experienced in the previous decade. If the trend established in the 1980s is extrapolated, then midway through the 1990s would see the underlying level down to 50,000 marriages per year, only two-thirds of the 1968 level.

The estimated cyclical pattern is constrained by the lack of intervention on this component at 1980 (recall that the 1980 intervention included a change in the trend component prior only). The cycle component routine discount factor, 0.9, does allow some movement but nothing dramatically sudden. The pattern at the end of the series is different from that at the beginning in two respects: the peak to trough variation (cycle amplitude) is much greater in the latter years; and the cycle undergoes a change of shape. The restriction in this analysis of a smoothly changing cycle may be inappropriate. Certainly there is a question mark over whether the structural change in 1980 affected the cycle pattern as well as the trend.

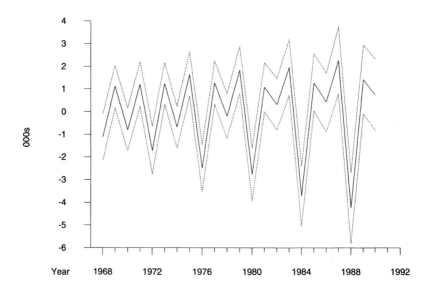

Figure 6.3 Estimated cyclical pattern from analysis I.

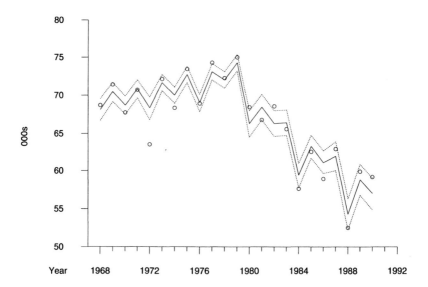

Figure 6.4 Estimated values from analysis I.

6.2 Analysis II

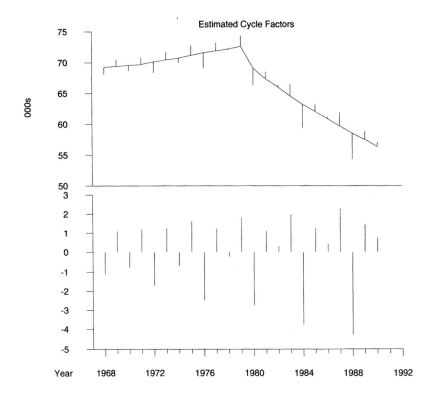

Figure 6.5 Alternative view of estimated cycle component.

With these identified shortcomings in the analysis it would be no great surprise to observe residual structure in the fitted errors. In fact, there is a large sample autocorrelation at lag six (exceeding the approximate 90% bound of $2/\sqrt{22} = 0.43$—remember that of the 23 data points, 1 was omitted as an outlier), as well as some large individual residuals.

6.2 Analysis II

In this second analysis we modify the interventions made at 1980. For the trend component the level prior is reset to 67 with standard deviation 1 and the growth prior is reset to -1.2 with standard deviation 0.5. These values are a tightening up of the intervention in the preceding analysis

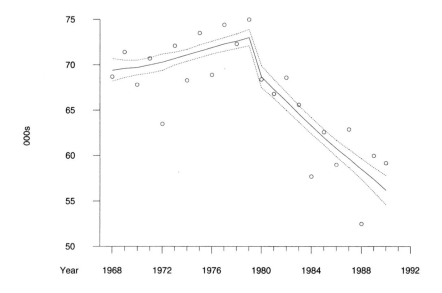

Figure 6.6 Estimated underlying level from analysis II.

based on the results obtained therefrom. We also reset the prior on the cycle component. Existing information about the pattern is discarded by setting all cycle factor priors to 0 with standard deviation 2. This allows the model to learn about a possibly quite different cycle pattern post-1980 from that estimated up to 1979. It does not force a new pattern however. The results of this analysis are shown in Figures 6.6–6.8.

Figure 6.6 shows the estimated trend. It is quite similar to that of the previous analysis (Figure 6.2), the present analysis—with a more informed intervention—giving a clearer view of the change from pre-1979 to post-1980.

More interesting are the estimates of the cycle component. Permitting the pattern to change radically at 1980 has made a significant difference to the analysis. The pre-1979 and post-1980 cycles quite clearly do have substantially different characteristics. The pattern in the first half is more tightly estimated than before, forming a very even 'N' shape. 1976 is estimated to be a particularly low year in this rising trend period. The cycle component in the latter years is less tightly estimated than in the early years: there is a lot of change exhibited. The most we can say from these estimates is that the four year trough becomes more pronounced, the intervening three years being fairly constant. The double peak/trough pattern of the first half of the series is no longer present.

6.2 Analysis II

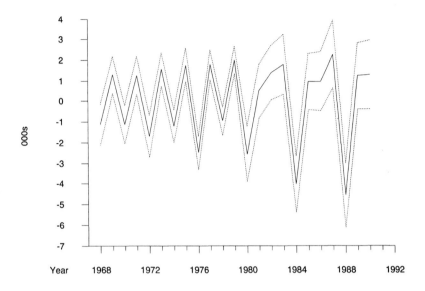

Figure 6.7 Estimated cyclical pattern from analysis II.

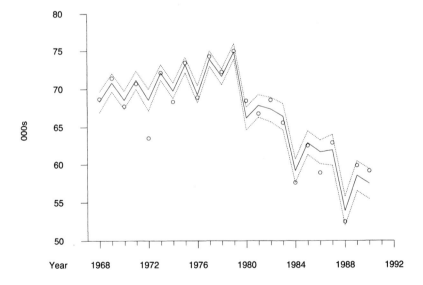

Figure 6.8 Estimated values from analysis II.

The last comment is perhaps premature. Overall fitted values in Figure 6.8 show that this analysis has probably not done a good job of identifying the cycle pattern post-1980. The problem is with the modelling of the trend. Rather than the steady linear decline pursued thus far, evidence here suggests the alternative of a series of sudden drops in 1984 and 1988, in addition to the drop in 1980 already modelled, may be more appropriate.

6.3 Analysis III

In this third analysis we investigate the alternative trend model of sharp drops in 1984 and 1988 in addition to the drop in 1980. Intervention at 1980 is similar to analysis II, only we tighten the growth prior standard deviation from 1 as used before to 0.5 here, following the results of the previous analysis. At 1984 we intervene to set the level prior to 61 with standard deviation 2, and growth prior to 0 with standard deviation 0.5. The cycle prior is set to be noninformative as before, with factor means 0 and standard deviation 2. A qualitatively similar intervention is made for 1988. The level prior is set to 55 (2), growth prior to 0 (0.5), and cycle factor priors to 0 (2). The level prior estimates selected for these interventions were obtained by inspecting the results of the previous analysis, always including a large degree of uncertainty.

The effect of the three interventions is to separate the series into four regions within which the model components are estimated with a degree of independence. The results of the analysis are shown in Figures 6.9 to 6.11. Looking at the estimated trend in Figure 6.9 the 'region separation' can be clearly discerned. The 1968–1979 trend is, as expected, what we have seen before. In each of the other three regions the trend is changed a little from the no-growth prior. In 1980–83 there is a very small decline, while in both 1984–87 and 1988–90 there is somewhat sharper growth (about the same rate as the 1968–79 period). However, these estimates are quite uncertain because of the very small number of data points in each case. A constant level—zero growth—trend (the prior) is certainly consistent with the data.

Turning attention to the cycle component in Figure 6.10 an interesting story unfolds. Before 1979 the plot is much the same as in the previous analysis. The one notable change is that the 1976 estimate no longer stands out from the other cycle troughs in 1968–79. The reason for this is the greater extent of the decoupling of the post-1979 data. The deep troughs in 1980, 1984, and 1988 no longer exert a 'pulling down' influence on 1976. Take a look at Figures 6.3, 6.7, and 6.10. Notice in Figure 6.3 how the cycle trough consistently becomes deeper with each cycle (Figure 6.5 shows this very clearly). The same effect is apparent in Figure 6.7 but less so because of the cycle component 'decoupling' intervention in 1980. As just remarked,

6.3 Analysis III

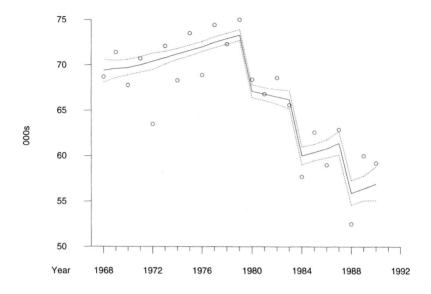

Figure 6.9 Estimated underlying level from analysis III.

the effect disappears entirely in the more complete decoupling analysis in Figure 6.10. (You have probably realised that the extent of the decoupling could be increased in analysis II by increasing the prior uncertainty at the 1980 intervention, a suggestion pursued in the exercises.)

The estimated cycle patterns for 1984–87 and 1988–90 are quite different from the patterns of the previous analysis. Clearly 1980–83 is something of a transitory phase. In the final years the amplitude increases to almost three times that experienced during 1968–79.

The fitted estimates in Figure 6.11 are quite precisely determined, and the fit to the data series is generally quite close (with the exception of the outlier in 1972). Certainly the comparison with the two previous analyses is very favourable. The residuals from this latest analysis no longer exhibit the strong lag six autocorrelation present in the previous analyses. The forecast residuals do give a visual impression of (negative) first order autocorrelation but the sample value is less that the 95% bound. For forecasting we could explore alternative discount factors to obviate the effect, but there would be no material change to the conclusions of the analysis here.

In Section 6.1 we remarked that if the post-1980 trend estimated there was extrapolated, then the underlying level marriages per year would fall to 50,000 by the mid-1990s. The analysis of this section suggests otherwise.

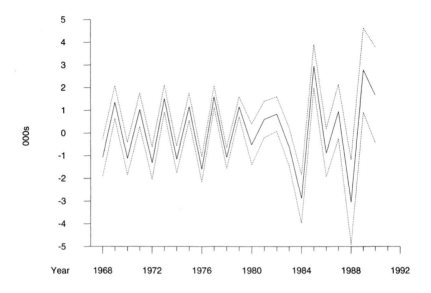

Figure 6.10 Estimated cyclical pattern from analysis III.

One final comment: The evidence for a level step change from 1987–88 is much less clear-cut than for the changes in 1979–80 and 1983–84. It is possible that the period from 1984 to 1990 is one of continual trend decline. This suggestion is pursued in the exercises.

6.4 Conclusions

The number of marriages in Greece has a strong four year cycle with the lowest point in the cycle coinciding with leap years. This reflects the strong Greek national superstition that it is unlucky to marry in a leap year.

The period 1968–79 is characterised by a steady growth in underlying level. The only unusual point during this period is 1972: the recorded number of marriages is very low for that period even by leap year standards. (See the following section for a counter-suggestion.)

The year 1980 could be another very low leap year or (as we assumed in the analysis) it could be the first year of a sharply reduced level. In either case the years 1981–83 do not follow the typical post-leap-year cyclical pattern. Economic uncertainty associated with the change of government in October 1981 (an event widely anticipated as early as 1980) lends anecdotal support to the suggestion of a level change in 1981.

6.5 1972: Case for the Defence

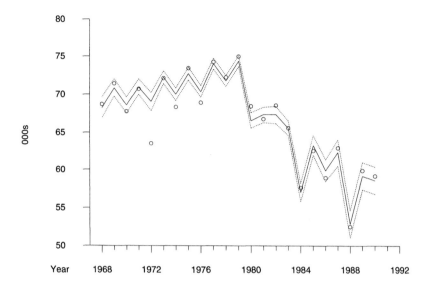

Figure 6.11 Estimated values from analysis III.

The year 1984 sees the start of a period characterised by a much reduced level, with possibly zero or slightly positive growth. The value in 1988 is very low, notwithstanding being a leap year. Evidence from the two succeeding years lends credence to the view that 1988 is yet another point of level reduction but this is not clear-cut. The cyclical pattern is consistent over 1968–79, and over 1984–91, although the changes around 1988 leave a question mark over possible cyclical change near that time. However, the patterns in these two periods have different characteristics. The overall decline from 1979 to 1991 may well be partly related to the performance of the Greek economy because marriage involves significant expense.

6.5 1972: Case for the Defence

Up to this point we have assumed that 1972 was an unusually bad year for marriages in Greece. The case has been presented only for the prosecution. It is the contention of the defence that 1972 does not represent an unusually low value and that, on the contrary, it is more representative of the true cycle behaviour than is 1976.

Consider the marriage totals recorded for the leap years 1972, 1980, 1984, and 1988. All of these years exhibit deeper troughs from the preceding and succeeding years than does 1976. Admittedly the 1984 and 1988 values are compounded by level reductions (it seems) but the magnitudes of the 1984–85 and 1988–89 differences are more in line with the 1972–73 value than the 1976–77 value. The outturn in 1980 may be compounded by a level decrease or not—the level decrease may have occurred in 1981—but it is impossible to say from the figures alone.

The defence is interesting and certainly raises a valid point. We leave it to you to pursue in the exercises.

6.6 Regional Analysis

We now turn up the magnification of our analysis microscope and look into the regional breakdown of the marriage series. There is much variation among individual regional patterns and some very striking contrasts. Official figures are provided for several mainland regions and three island groupings. The mainland regions included in the figures are Athens, Macedonia, Epirus, Peloponnesos, Thessaly, Thrace, and 'the rest'. The many Greek islands are grouped as Crete, Ionian islands, and Aegean islands.

Our analysis is restricted to a brief commentary. We leave the formal analysis and estimation as an exercise for you.

6.6.1 Athens, Macedonia

Figure 6.12 shows the marriage series for Athens and Macedonia. Both series exhibit overall patterns quite similar to the pattern of the full series (compare Figure 6.11). About one-third of the population of Greece lives in the capital city, a fact clearly reflected in the Athens marriage series. Interestingly, growth in the 1968–79 period is stronger than for the all-Greece series. The region of Macedonia (not to be confused with the now independent republic of the former Yugoslavia) covers central and northern Greece. The marriage series is very similar to the Athens data, the major difference being a decline over the early years.

6.6.2 Epirus, Peloponnesos, Thessaly, Thrace

Figure 6.13 shows the marriage series for Epirus, Peloponnesos, Thessaly, and Thrace. These four regions cover western, southern, central, and northeastern Greece respectively. The pattern of these marriage series is different from those seen previously. The 1968–79 period is flat and the subsequent period is characterised by continual decline rather than a series of sharp level changes. There is almost no consistent cyclical variation.

6.6 Regional Analysis

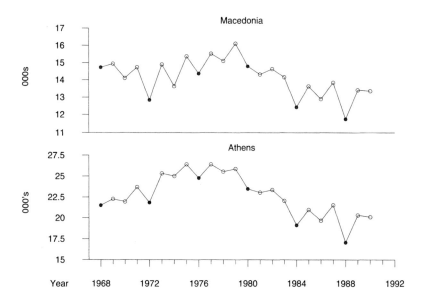

Figure 6.12 Annual number of marriages in Athens and Macedonia (thousands).

6.6.3 Rest of Mainland

Figure 6.14 shows the marriage series for the remainder of mainland Greece. The general trend is one of more or less continual decline with 1973–82 being flat. There is an indication of a sharp reduction in underlying level after 1982. Cyclical variation clearly exists but it is not very consistent.

6.6.4 Crete, Aegean Islands, Ionian Islands

Figure 6.15 shows the marriage series for Crete, the Aegean islands, and the Ionian islands. Crete is the large island off the southern tip of mainland Greece. Superficially the series behaviour is qualitatively similar to the all-Greece series but on a much smaller scale of course. Closer inspection suggests that a more careful characterisation would describe 1968–72 as a period of decreasing underlying level, being followed by a rising underlying level over 1973–80.

The Ionian islands (off western mainland Greece) series exhibits a pattern of consistent decline together with the strong four year cycle. The year 1982 is unusual in that it does not conform to the cyclical pattern (a low) present almost everywhere else in the series. Just how unexpectedly high

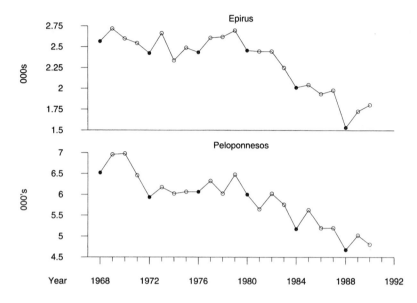

Figure 6.13a Annual number of marriages in Epirus and Peloponnesos (thousands).

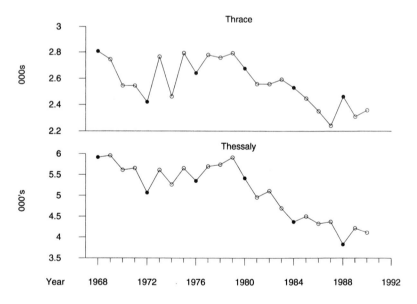

Figure 6.13b Annual number of marriages in Thessaly and Thrace (thousands).

6.7 Summary

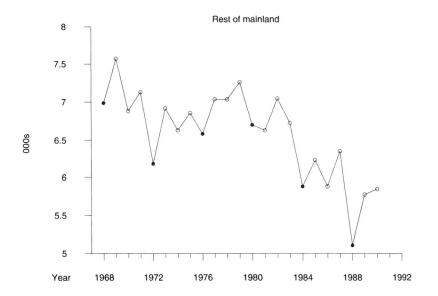

Figure 6.14 Annual number of marriages in the remainder of mainland Greece (thousands).

the 1982 value was can be seen from Figure 6.16 which shows the estimated values for 1972–90. The analysis here assumes smoothly changing trend and seasonal components (discount factors 0.95 and 0.9 respectively). There are no interventions for structural change. The value for 1982 was treated as an outlier so that it would not distort the estimated cyclical pattern present in the rest of the series. The first four years (1968–71) were omitted for the same reason.

The estimated model certainly fits the data quite well, although there is some evidence of greater movement in the underlying level than allowed for in this analysis. Notwithstanding that, it is clear that the number of weddings in the Ionian islands in 1982 was about 80 (40–120) more than expected given the four year pattern prevalent over 1972–90.

The Aegean islands (off eastern mainland Greece) series also exhibits a general pattern of decline. Both the trend and the cyclical variation are more erratic than for the other series in this group.

The consistently declining trend in marriages for the small islands series contrasts with the flat or growing trend in the other regions over the years 1968–72. This effect may well be an indication of a gradual migration of the islands' populations to the mainland or abroad.

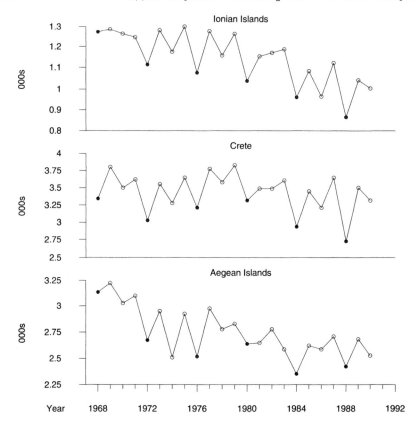

Figure 6.15 Annual number of marriages in Crete, the Aegean islands, and the Ionian islands (thousands).

6.7 Summary

In this chapter we analysed numbers of marriages in Greece for the period 1968 to 1990. The series is interesting for a pronounced four year cycle apparently driven by national superstition. Other interesting features of the series are sudden and dramatic drops in underlying level, and the overall decline after 1980 following a decade of steady increase.

We also briefly examined a regional breakdown of the national picture. The story deepened as we discovered several different patterns in the regional series. In particular, several regions show a decline over the first half of the period while nationally there was steady underlying growth.

We end this chapter with a personal note. One of us (AP) contributed to the Athens marriage total for 1993.

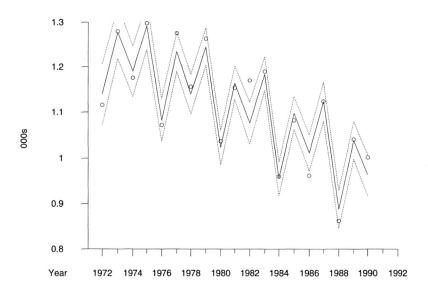

Figure 6.16 Estimated number of weddings for the Ionian islands (thousands).

6.8 Exercises

Exercise 6.1 The wedding of the Crown Prince of Japan took place in May 1993. In early 1992 there was speculation in the press that this event would impact the Japanese economy through:

(i) increased demand for televisions and video recorders;
(ii) an increase in the number of marriages in 1993 and 1994.

Obtain relevant data series (you will need several years of history as well as the figures for 1993/4) and determine whether the speculation was borne out.

Exercise 6.2 Compare results from the analysis in Section 6.1 with a similar analysis that does not exclude the value for 1972. Comment on the differences.

Exercise 6.3 Refer to Section 6.2. The covariance matrix of the seasonal component prior following the intervention at 1980 is the diagonal matrix $4I$. Apply the zero sum constraint (see subsection 3.5.2 of Chapter 3 and subsection 10.11.4 of Chapter 10) to obtain the actual prior covariance matrix implied by the intervention. Note that BATS rescales the constrained covariance matrix to preserve the total variance (the trace) of the input specification.

Repeat analysis II with greater prior uncertainty on the cycle component: experiment with several values. Compare the estimated seasonal pattern with the estimates in the text Figures 6.3, 6.7, and 6.10.

Exercise 6.4 Refer to Section 6.3. Repeat analysis III but change the level component prior at the 1984 and 1988 interventions to higher mean values than used in the text. What do you notice about the estimated values compared with those from the analysis in the text?

Exercise 6.5 Refer to Section 6.3. Experiment with changing the component routine discount factors (use the same interventions as in the text). Examine the forecast and fitted residual sample autocorrelations. Comment on the claim in the text that changing discount factors by a small amount does not materially alter the conclusions of the analysis. What happens to the sample autocorrelations of the forecast residuals?

Exercise 6.6 The analyses in the text use routine component discount factors of 0.95 for the trend and 0.9 for the cycle. With the structural changes identified and modelled with appropriate interventions there is indication that the trend discount may be usefully increased. It may also be useful to increase the dynamic movement allowed for the cycle component by lowering its discount factor. Following the analysis in Section 6.3, experiment.

Exercise 6.7 Refer to Section 6.5. Consider the argument there and make an appropriate analysis. Pay careful attention to alternative discount factors. From your results estimate how unusually high the 1976 value is. What does your analysis indicate about the value for 1968?

Exercise 6.8 It is clear that some people go for an early marriage—the year before a leap year—and some go for a deferred marriage—postponing nuptials until the year after a leap year. A 'direct' model for this description consists of a trend—represented by every fourth year two years shifted from leap years—and adjustments from this trend for the leap year (negative) and the two leap-adjacent years (positive):

$$\text{marriages}_t = \text{trend}_t + \alpha_t \text{leap}_t + \beta_t \text{leap}_{t-1} + \gamma_t \text{leap}_{t+1},$$

where $leap_t$ is an indicator variable, 1 in a leap year, 0 otherwise. This model is more restrictive than the trend/cycle model analysed in the text: the non-leap-adjacent year being assumed to be at the underlying level in the present case. Explore this model and compare your results with those in the text.

How could you analyse this model using the trend/cycle model? [**Hint** Think about the prior specification for the cycle factors.]

Chapter 7

Further Examples and Exercises

In this chapter we present several data sets with brief descriptions, preliminary analysis, and suggestions for further work. Many additional data sets are illustrated without comment. They show the diversity of subject areas where time series arise, from medicine to astronomy, geology to finance, water quality to commerce, speech therapy to agriculture, biology to travel, and many more. All of the data sets are included on the BATS program diskette.

7.1 Nile River Volume

Figure 7.1 shows the annual volume of the Nile River for the years 1871 to 1970 taken from Cobb (1978). Balke (1993) discusses this series to demonstrate a procedure for detecting and modelling level shifts in ARIMA models. Balke states, "This series was examined by Cobb (1978), Carlstein (1988), and Dümbgen (1991), and these authors suggested that the Nile River volume experienced a permanent decline in the mean of its distribution in 1899."

Using his procedure Balke identified the following first order autoregressive model with a level shift in 1899 (Balke, 1993, p. 82),

$$Y_t = 248 \text{LS}1899_t + [1/(1 - 0.161B)][920.7 + a_t],$$
$$\hat{\sigma}_a = 127.3$$

where Y_t is the Nile volume (scaled by a factor of 10E8), LS1899 is a dummy variable for indicating the level shift in 1899 (so that $\text{LS}1899_t = 1$

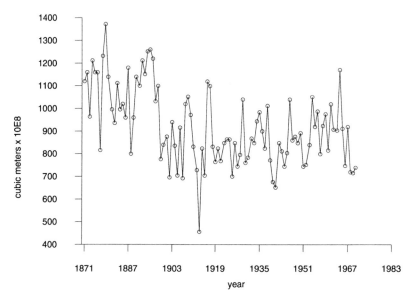

Figure 7.1 Annual volume of the Nile River.

for $t \geq 1899$ and 0 otherwise), and B is the back shift operator (so that $B^i X_t = X_{t-i}$). The estimated level in Balke's model is therefore given by $920.7 \times (1 - 0.161) = 1097$ before 1899, and $1097 - 248 = 849$ in 1899 and thereafter.

The autoregressive coefficient estimate, 0.161, has an estimated standard deviation of 0.101 and would therefore usually be considered not significant, reducing the model to a constant with a shift in 1899. Balke retains the AR model, presumably because his methodology is designed for detecting/estimating level shifts in ARIMA models. As will be shown below, there is no real evidence for the autoregressive structure.

The first order polynomial DLM for the Nile River volume has the simple form
$$\text{volume}_t = \text{level}_t + \nu_t, \qquad \nu_t \sim N[0, V_t],$$
$$\text{level}_t = \text{level}_{t-1} + \Delta\text{level}_t, \qquad \Delta\text{level}_t \sim N[0, W_t].$$

The static model is the special case where the system equation innovation variance W_t is zero (equivalent to a level component unit discount factor), so that the system equation reduces to
$$\text{level}_t = \text{level}_{t-1}.$$

On-line analysis of the static level model with a variance intervention at 1899 is illustrated in Figure 7.2. Very vague initial priors were specified as

7.1 Nile River Volume

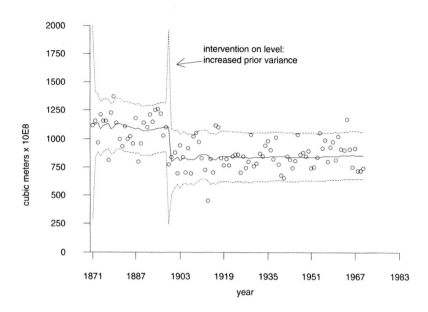

Figure 7.2 One step ahead forecasts and 90% uncertainty limits for the Nile River volume; forward intervention on level uncertainty for 1899.

is clearly indicated by the 90% uncertainty limits. The level prior variance for 1899 was increased from the estimated value of 28 to 500 to allow for the possibility of change. The prior mean was unchanged from its estimated value of 1098 so the issue of change is not prejudged. The forecast for 1899 is at the level estimated from the data up to 1898 since the forecast function for the model is just the level, but the uncertainty is greatly increased as a result of the intervention. When the 'low' value for 1899 occurs, the level estimate is quickly adjusted downward—the movement is allowed by the now very uncertain prior. Subsequent values in the series remain around the lower level consolidating the new level estimate, quickly reducing uncertainty.

The retrospectively estimated level is shown in Figure 7.3. The form of the model is a constant and the analysis specified no dynamic movement in the level, hence the straight line form of the estimates. Of course, the intervention in 1899 allowed for a once only change in the value of the level at that time—but did not impose such a change—and the data says that there is a change. The pre-1899 level is estimated at 1084 and the 1899 and after level at 850, with residual standard deviation at 126. These estimates are essentially identical to those from Balke's AR(1) model.

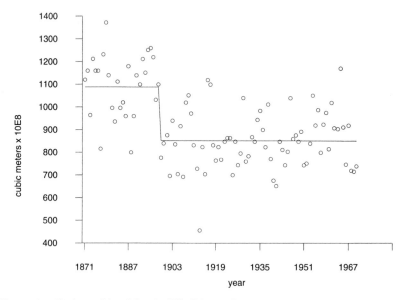

Figure 7.3 Estimated level for the Nile River volume.

None of the sample autocorrelations of the estimated residuals, shown in Figure 7.4, exceed the approximate two standard deviation limit ($2 \times 100^{-1/2} = 0.2$) which suggests that there is no autocorrelation structure to model. This supports the comment we made above about the autoregressive parameter estimate in Balke's model.

This brief analysis has demonstrated how a static model is analysed in the dynamic model framework. One of the benefits of this formulation and analysis is the straightforward production of forecasts as they would be done in real time (Figure 7.2). The analysis has also illustrated how component shifts are handled directly by estimating shifts in the component itself. It is not necessary to introduce artificial components or dummy variables to model component changes as is the case with ARIMA models.

Analysis Suggestions

The level change point, 1899, was flagged as a potential time of change by the BATS forecast monitoring system. Repeat our analysis and identify other potential points of change.

In the Nile volume series the value for the year 1911 is less than half the average post-1899 level and the lowest figure in the series by a considerable margin. Research this matter and try to discover the circumstances that caused the small value. Your starting point should be the original data: is

7.1 Nile River Volume

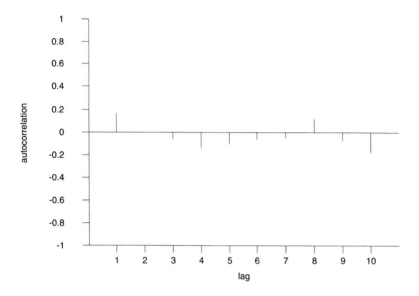

Figure 7.4 Autocorrelation of estimated residuals for the first order polynomial static DLM analysis of the Nile River volume.

it possible that a transcription error was made? Are there any other points in the series that you consider interesting, individually or as a group?

The Aswan Dam on the Nile was opened in 1955. Investigate whether there has been any associated change in the volume of the Nile River. (You may want to collect data since 1970 to get a clearer picture.) Begin your investigation with a dynamic analysis of the first order polynomial DLM (including forecast monitoring); experiment with alternative discount factors for the level.

The Nile volume in 1899, 774E8 cubic meters, is not unusually low in comparison with the period 1871 to 1888—in that period there were two other such 'low' volumes recorded, 813E8 in 1877 and 799E8 in 1888. It is not unreasonable to suppose, therefore, that the reduction in volume actually occurred in 1900 or even 1901, and that 1899 (and possibly 1900) happened to be low values by chance but belonging to the pre-drop group. Balke (1993) cites Cobb (1978) as discussing independent evidence on tropical rainfall in support of the Nile volume reduction. Examine the evidence and decide when you believe the level change actually occurred.

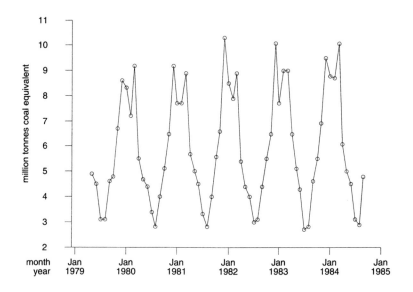

Figure 7.5 Monthly inland natural gas consumption in the U.K. (Source: Central Statistical Office Monthly Digest.)

7.2 Gas Consumption

The data in Figure 7.5 is monthly consumption of natural gas in the United Kingdom for May 1979 to September 1984. The series has a very pronounced seasonal pattern, rising to a peak in the winter months and falling to a minimum in the summer. The underlying level appears to be approximately constant.

Analysis Suggestions

Analyse the gas consumption series and determine a suitably parsimonious description of the seasonal pattern. Is a full harmonic description necessary? In particular, examine a reduced form seasonal component comprising just the fundamental and fourth harmonics, representing annual and quarterly cycles respectively. Discuss the stability of the estimated seasonal pattern. Examine the underlying trend for the series. Is there a long term pattern of growth? Examine summary forecast statistics (mean square forecast error, mean absolute forecast error, forecast log-likelihood) for a range of seasonal and trend component discount factors. What do you conclude about the suitability of a static model for this series? Do you detect any evidence of movement in the underlying level of gas consump-

7.2 Gas Consumption

tion? Using your selected model project gas consumption for 1985. Obtain actual consumption figures from the CSO Monthly Digest and assess the performance of your forecasts.

A national strike by the majority of Britain's coal miners began in 1984 and continued for over a year. Initially coal stocks were drawn upon to make up for the loss in production but soon alternative energy sources began to be substituted for coal. The year 1984 and subsequent years were a time of massive change in the U.K. energy sector. Obtain figures for the major energy sources for the period and research the effects.

An annual cycle is wholly expected in the demand of a commodity like gas because of its use for heating and the pronounced annual temperature cycle in the U.K. What factors might be responsible for the higher frequency cycles observed? (Among other factors, recall the discussion in Chapter 3 on harmonic analysis and asymmetric cycles.)

The national company responsible for supplying gas in the U.K. was sold by the State into private ownership in 1986. Obtain and analyse post-1984 data: do you detect any evidence of changes in consumption patterns following the privatisation? Obtain retail gas price information and explore the relationship between consumption and price: is there a discernible price/consumption relationship? Contrast this relationship for the periods before and after privatisation.

Electricity is a major competitor to gas for heating, lighting, and cooking. The electricity supply and generation industries were privatised in 1991. Is there any evidence of a change in gas consumption corresponding to this event?

The U.K. economy went into the deepest recession of the twentieth century in 1990, and only began to emerge in the middle of 1993. Is this reflected in gas consumption?

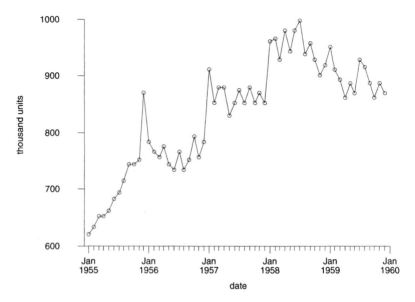

Figure 7.6 Monthly sales tobacco.

7.3 Retail Sales

Figure 7.6 shows the monthly sales of a tobacco company over a period of five years from January 1955 to December 1959. The series is interesting because of the variety of non-standard behaviour it exhibits. Unusual, outlying points, abrupt level changes, and changes in general variability are all present.

Analysis Suggestions

Using a second order polynomial trend (linear growth) DLM with discount factor 0.9 compute the one step ahead forecasts (use vague initial priors). Calculate the retrospectively fitted values and comment upon the way in which these values 'fit' the data.

Repeat the previous analysis this time using a monitor scheme of the kind described in Chapter 3. Select all three monitors, variance inflation, level increase, and level decrease, in BATS. Use 'ignorance' intervention at each monitor signal of forecast breakdown. (This is the 'automatic' option in BATS.) Compare the forecasts and fitted values for this monitoring/intervention analysis with the corresponding values from the non-intervention analysis previously. Comment upon the changes signalled by the monitor.

7.3 Retail Sales

Experiment with alternative discount factors, including the static value 1.0, for the trend component in both the intervention and nonintervention analyses. How do estimates of the level changes in 1956 and 1957 vary with different discount factors?

Describe the pattern of observation noise in the series. Is a variance power law suggested?

What kinds of commercial activity might result in the abrupt increases in sales volume?

The tobacco series is analysed in Harrison and Stevens (1976).

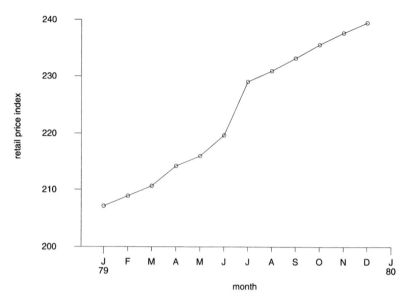

Figure 7.7 United Kingdom retail price index.

7.4 Inflation

Figure 7.7 shows the United Kingdom retail price index for the 12 months of 1979. The series is clearly linear, with very little noise, and a definite jump in the level in July. One month earlier it was announced by the Chancellor of the Exchequer (finance minister) that valued added tax (VAT) was to be increased by 10% to 15%.

Analysis Suggestions

Analyse the series for the first six months of 1979 and make a forecast for July. Assuming that VAT is applied to almost all of the elements of the retail price index, what effect would you expect the VAT increase to have on the index? Perform this intervention analysis. How well did you do?

Economic variables typically exhibit long term growth patterns that are reasonably constant in percentage terms (that is, exponential growth). When analysing such series would a variance law be appropriate? What about a log transformation?

The U.K. retail price index is analysed in Harrison (1988).

7.5 United Kingdom Marriages

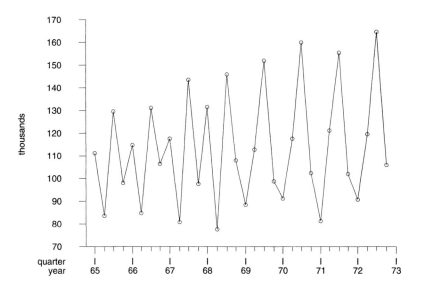

Figure 7.8 Quarterly marriages in the United Kingdom. (Source: CSO Monthly Digest of Statistics.)

7.5 United Kingdom Marriages

Figure 7.8 shows the number of marriages each quarter in the United Kingdom for the period January 1965 to December 1970. This series is characterised by an underlying level which is consistently increasing, and a strong seasonal pattern which undergoes a change of form in 1968.

Analysis Suggestions

Analyse the data up to the end of 1968 and make projections for 1969. Compare your forecasts with the recorded values for that year. Notice how the seasonal pattern has changed.

Perform an intervention at the first quarter of 1969 and increase the prior variances on the seasonal component (do not alter the trend component prior). Examine the retrospectively fitted seasonal pattern: notice how the intervention allows the model to estimate the seasonal pattern in both its pre- and post-1969 forms.

In the first quarter of 1968 the Chancellor of the Exchequer announced in his budget that the tax basis for married couples was to be changed. The change was to remove the then existing entitlement of newlyweds to reclaim certain taxes paid in the year in which they married. Before the change the U.K. tax system encouraged couples to marry in the first quarter of the year when the tax refund was maximised, and discouraged marriage in the second quarter when the benefit was minimised. That explains the winter seasonal peak in the marriage series in the pre-1969 period. Of course, the climate effect is responsible for the summer peak. Following the change in tax regime there was no longer any financial advantage to be gained by marrying in the winter. With the climate being a disincentive it might be expected that the result of the change would be the removal of the winter seasonal peak from the series. This is indeed what happened as the data and your previous analysis show.

Since the tax changes were announced well in advance it was possible at the time to adjust forecasts for 1969. From your first forecasts for 1969 and the discussion in the previous paragraph, revise your forecasts. Perform an intervention analysis, altering the seasonal component prior in line with your revised forecasts. Compare the forecasts from this intervention analysis with the forecasts from the variance-only intervention performed previously.

Did the tax changes have any effect on the underlying trend of numbers of marriages?

7.6 Housing Starts

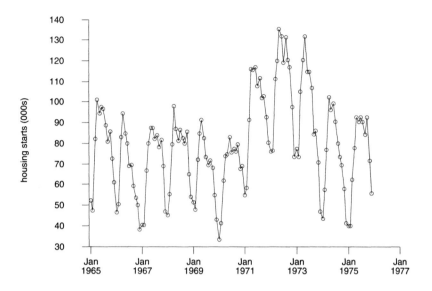

Figure 7.9 Monthly United States housing starts of single-family structures.

7.6 Housing Starts

Figure 7.9 shows the monthly United States starts of privately owned single-family structures from January 1965 to December 1975 (Abraham and Ledolter, 1983). The series exhibits a rather noisy annual cycle combined with a generally constant underlying level. However, there is a precipitous jump in the level early in 1971 and a subsequent equally precipitous drop in 1973.

Analysis Suggestions

Investigate a first order polynomial trend and unrestricted seasonal component DLM for this series. Use the forecast monitoring system in BATS to determine at which months the level changes occurred. Estimate the magnitude of the changes. Does the series return to the pre-1971 level after the 1973 drop?

By examining the estimated harmonics of the seasonal component determine if a restricted harmonic pattern is appropriate for the series. Is there any evidence of fundamental change in the seasonal pattern, either associated with the level changes or otherwise?

Abraham and Ledolter (1983) fit a first order multiplicative ARIMA model to the housing starts data for 1965 to 1974 giving forecasts for 1975 as in

Table 7.1 Multiplicative ARIMA forecasts for housing starts[1]

Time	Forecast	Actual
1975/1	37.06 (6.51)	39.791
1975/2	40.25 (8.04)	39.959
1975/3	68.21 (9.32)	62.498
1975/4	84.61(10.44)	77.777
1975/5	86.29(11.46)	92.782
1975/6	83.44(12.39)	90.284
1975/7	77.29(13.26)	92.782
1975/8	75.60(14.07)	90.655
1975/9	67.63(14.84)	84.517
1975/10	67.64(15.57)	93.826
1975/11	54.30(16.27)	71.646
1975/12	39.29(16.94)	55.650

[1] Forecast origin is December 1974.

Table 7.1. Compare these forecasts with forecasts from your selected DLM. (Remember that forecasts for 1975 should be made from the standpoint of end-1974.) What do you conclude?

The same series is analysed in detail in Pankratz (1991). He identifies the same multiplicative ARIMA model as Abraham and Ledolter but reports slightly different parameter estimates and forecasts. Pankratz also analyses the corresponding series of housing sales and the relationship between starts and sales, with sales driving starts. In his analysis the pure time series model produces better point predictions but with wider uncertainty limits than the regression model. Construct a DLM for housing starts with sales as an explanatory variable: is a lagged relationship suggested? If starts and sales are contemporaneously related, how useful is the regression model for prediction? The housing starts and sales series are shown together in Figure 7.10.

The 1971 expansion in sales of housing is quite likely a result of the massive explosion in money supply (easy credit, etc.) at that time. The contraction of sales in 1973 may well be related: the catching up of inflation induced by the earlier money supply increase. Obtain money supply and credit figures for the period and investigate these conjectures.

7.6 Housing Starts

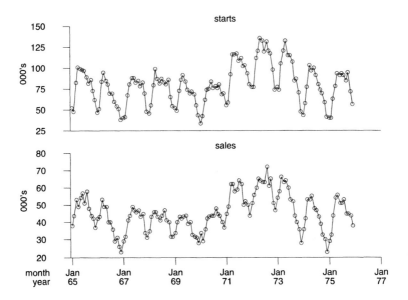

Figure 7.10 Monthly United States housing starts and sales of single-family structures.

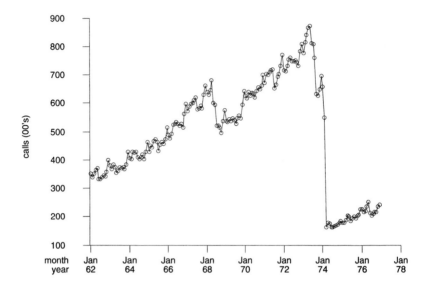

Figure 7.11 Monthly average daily calls to Cincinnati directory assistance.

7.7 Telephone Calls

Figure 7.11 shows the average number of calls per day in each month to Cincinnati directory assistance (Pankratz, 1991, Series 28). The series is striking for three sharp drops, including one quite catastrophic drop. These structural adjustments aside, the series shows a remarkably persistent and consistent growth pattern.

It is clear from the figure that the sudden drops in level take place over a span of several months. The first, in early 1968, proceeds over three months; the second, in the middle of 1973, is slightly longer lived, spanning four months; the third (coming fast on the heals of the second) is by far the greatest shift and occurs over the shortest time span: just two months.

Analysis Suggestions

Cincinatti Bell (the local telephone services provider) introduced a 20 cents per-call charge for directory assistance in March 1974 (Pankratz, 1991, p. 276). Can you find reasons for the previous two sharp drops in the series? The February figure for 1974, the month preceding the introduction of the call charge, is a considerable reduction from the previously established trend. Can you explain this?

7.7 Telephone Calls

Using a second order polynomial trend DLM estimate the rate of increase in the series for the three drop-separated periods. How do these rates compare? Do you detect any evidence of higher order growth (a quadratic or exponential trend for example)?

Pankratz (1991, p. 276), suggests analysing this series after performing a logarithmic transformation. Do you consider this necessary? If so, why? Compare forecasts from analysing the log transformed series with forecasts for the raw series. What do you conclude? Investigate the usefulness of a power law when modelling the untransformed series.

Do you detect any evidence of seasonal behaviour in the series?

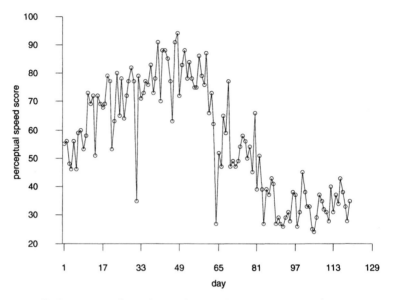

Figure 7.12 Daily perceptual speed score for a schizophrenic patient.

7.8 Perceptual Speed

Figure 7.12 shows the score achieved on a test of perception skills by a patient diagnosed as schizophrenic (Pankratz, 1991, Series 29). The patient began a course of drug treatment on day 61. The drug, chlorpromazine, is expected to reduce perceptual speed (Pankratz, 1991, p. 277).

Analysis Suggestions
Observation 31 seems to be inexplicably low. Lacking any information pertaining to this value it is reasonable to exclude it from analysis of the series. Observation 64 is also out of line with the surrounding values. Can it also be reasonably excluded?

When undergoing a series of tests designed to examine performance in a particular skill, it is often the case that there is an initial phase where general skill level at the task increases with repeated use. After this learning phase, one arrives at a plateau with further improvement requiring directed practice. Fit a second order polynomial trend model to the series and determine if this model of behaviour is a reasonable description of the speed score series.

It is of interest to determine the nature of the effect of the administered drug on perceptual performance. Is the effect immediate? Is it cumulative?

7.8 Perceptual Speed

Is there a gradual decline in performance with achievement tailing off to a stable lower level? Over what length of time are adjustments noticed? Are there any changes in the variability of the patient's response?

Regarding observation 64 as an outlier, fit the linear growth model both with and without an intervention at time 61. Characterise the perceptual ability of the patient throughout the experiment. How would you answer the questions posed in the previous paragraph?

Are there any grounds for supposing that the low value recorded for time 64 might be unrelated to chlorpromazine administration?

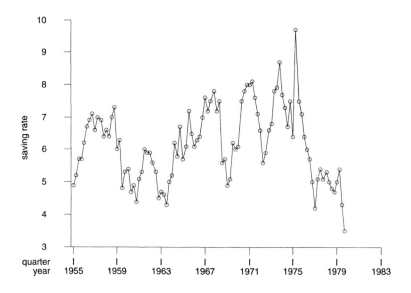

Figure 7.13 Quarterly U.S. saving rate.

7.9 Savings

Figure 7.13 shows the quarterly savings rate (percentage of income) in the United States from the first quarter of 1955 to the fourth quarter of 1979 (Pankratz, 1991, Series 3). In the second quarter of 1975 the U.S. Congress passed a law granting a one time tax rebate (Pankratz, 1991, p. 4). The permanent income hypothesis of economic theory says that such a tax rebate should result in a temporary rise in savings rate.

Analysis Suggestions

The savings rate series exhibits a good deal of variation. There is no evidence of annual seasonal movement, but there is a visual impression of a more long term cycle with a period varying between four and seven years. This could well be related to the business cycle of the U.S. economy. Obtain figures for U.S. national income for the period of the data and test that hypothesis.

Figure 7.14 shows the sample autocorrelations for the saving rate, showing clear evidence of a first order autoregressive relationship. Estimate a DLM comprising a level component and a regression on the one period lagged rate. Compare forecasting performance for a range of discount factors; in each case examine the autocorrelation structure of the forecast residuals.

7.9 Savings

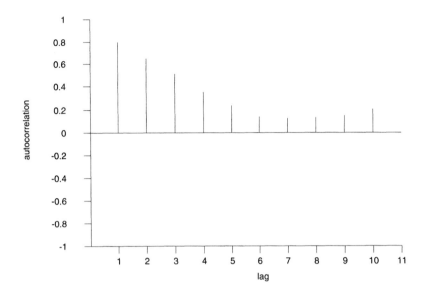

Figure 7.14 Sample autocorrelations of quarterly U.S. saving rate.

Is a static model adequate? Estimate the effect of the tax rebate on savings rate. What can you say about the trend in savings rate after 1975?

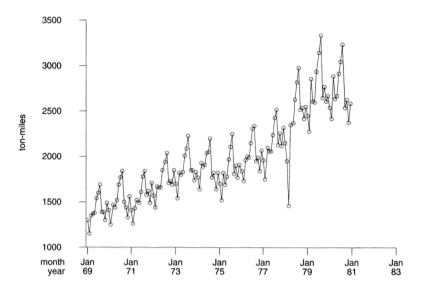

Figure 7.15 Air carrier freight volume: monthly.

7.10 Air Freight

Figure 7.15 shows the volume of freight (in ton-miles) carried by U.S. air carriers from January 1969 to December 1980. The general pattern of the data is one of underlying positive growth composed with a strong seasonal cycle. There is evidence of a reduction in the underlying level in 1974 which does not appear to be accompanied by a change in either the annual cycle or the long term growth. There is a subsequent increase in underlying level just four years later, preceded by an extremely low value in March 1983. It is also evident that the peak to trough seasonal amplitude increases in the later years.

We want to be more precise about the description. Exactly how stable is the seasonal variation? Is the underlying growth as we described or has our eyeballing missed some important details? Is March 1983 the only unusual point in the series? And, turning the focus to prediction, what is the likely outlook for 1981?

Figures 7.16 and 7.17 shows two alternative views of the deseasonalised freight series. Figure 7.16 shows the annual aggregate freight volume, while Figure 7.16 shows a centred 13 month moving average. The story here confirms the comments made above about the underlying trend, and in fact makes the picture even clearer. The two years 1973 and 1974 were a period

7.10 Air Freight

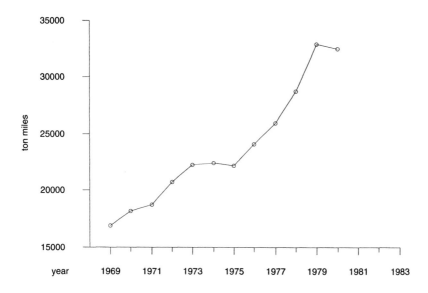

Figure 7.16 Air carrier freight volume: annual total.

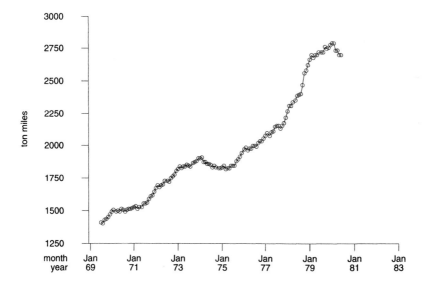

Figure 7.17 Air carrier freight volume: 13 month centred moving average.

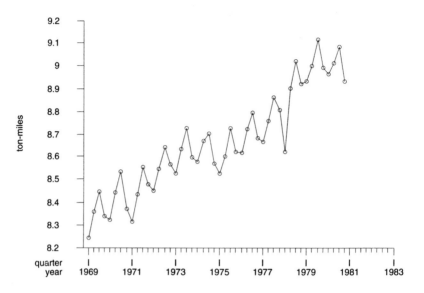

Figure 7.18 Air carrier freight volume: quarterly.

of zero growth, preceded by strong year-on-year growth, and followed by four years of even stronger growth. The final year of the data, 1980, once again breaks with the growth pattern, actually recording a slight decline from the previous year. It is interesting to note that the low March 1983 figure is not apparent in the annual series implying that it was offset by higher volumes later in that year.

Figure 7.18 shows the air freight volume by quarter, transformed to the logarithmic scale. The same general features as described for the monthly series are visible. In addition, the quarterly seasonal pattern can be seen to be quite consistent: starting low in the first quarter, rising in the second quarter, rising still higher in the third quarter, and dropping back in the fourth quarter. This triangular pattern is prevalent throughout the entire series. Notice that the transformation has stabilised the peak to trough range: it no longer increases with the level.

Figures 7.19 and 7.20 show the first differences of the freight series on the natural measurement scale and the log scale. On the natural scale the values for 3/78, 9/79, and 9/80 (and perhaps 3/79 also) show up as unusual by comparison with the other series values. On the log scale the values for 2/78 and 3/78 show up as unusual.

7.10 Air Freight 189

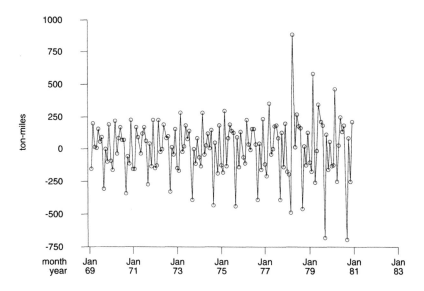

Figure 7.19 First differences of monthly air carrier freight volume.

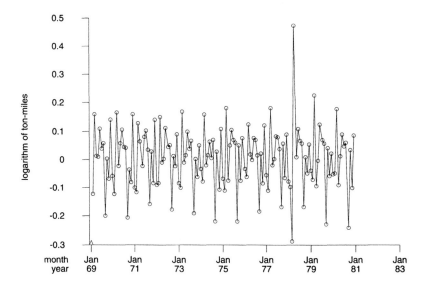

Figure 7.20 First differences of log monthly air carrier freight volume.

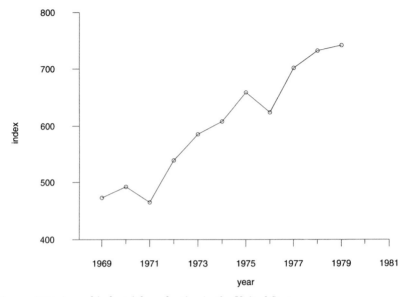

Figure 7.21 Annual industrial production in the United States.

Analysis Suggestions

Research the cause of the low freight volume recorded for March 1983. Investigate original sources for the data: is it possible that a transcription error was made? What would that imply for overall volume in 1983 (assuming that the other figures are correct)? Was there an industrial dispute that caused disruption of air-shipping? Can you determine anything about the other identified unusual points?

Investigate the state of the U.S. economy over the period covered by the series. Is there a relationship between gross domestic product and freight volume? Figures for annual U.S. industrial production (U.S. Department of Commerce) are shown in Figure 7.21; you may also want to obtain figures for total economic activity: why? (What problems would you encounter in attempting to consider such questions with more frequent observations, say monthly? Monthly figures for U.S. industrial production are included on the BATS diskette.)

Obtain figures for U.S. economic activity for 1982. On the basis of your conclusions regarding the relationship of economic activity to air freight volume, forecast the latter for 1982 on both monthly and quarterly time scales. Assess your forecasts against the outcome. How well did you do? Compare the quarterly point forecasts with quarterly aggregated monthly point forecasts.

7.10 Air Freight

The member countries of the oil producers cartel, OPEC, quadrupled the price of crude oil in 1973/74. Was that decision related in anyway to the cessation of growth in air freight in 1974 and 1975?

Analyse the monthly data on the logarithm transformed scale. Is the monthly seasonal pattern as stable as the quarterly pattern? How does the estimated underlying trend compare for the monthly and quarterly series? From the monthly analysis identify the likely turning points in the series: Does the quarterly aggregation obscure the timing?

Investigate alternative variance stabilising transformations, the square root transformation for example (see Chapter 4). Is the log scale the most appropriate?

West and Harrison (1989) describe an alternative model for series like the freight volume data where seasonal amplitude increases with level, the multiplicative seasonal model. The model is nonlinear but may be analysed in the linear framework by using a linearisation approximation. Such an approach offers the advantage of modelling a series on its natural scale, as opposed to taking transformations. The multiplicative model may be better than working with log transformed data in the present example because the underlying trend is clearly linear in the original metric. That linearity is slightly obscured on the log scale. Following West and Harrison (1989) write a computer program (or obtain the source code to BATS from the authors and extend that) to implement the multiplicative seasonal DLM. Does the multiplicative model outperform the linear model/log-transformation combination?

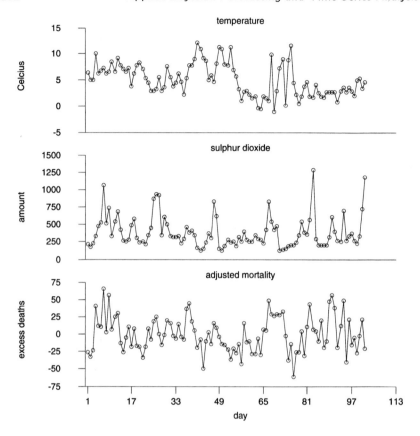

Figure 7.22 Daily London mortality (adjusted), sulphur dioxide, and temperature.

7.11 London Mortality

Figure 7.22 shows daily deaths, sulphur dioxide, and average temperature for the Winter of 1958 (Shumway, 1988). The deaths series has been adjusted (Shumway, 1988, p. 33) to remove the effect of a flu epidemic.

Analysis Suggestions

Go back to the original reference in which the mortality data was analysed, Shumway *et al.* (1983), and obtain the raw death figures. Fit a second order polynomial DLM (linear growth) model to estimate the underlying trend. What discount factor do you consider appropriate? Compare the residuals from the fitted model with the adjusted series shown here. Repeat the exercise using a third order polynomial (quadratic growth) trend. Are there

7.11 London Mortality

any significant differences in the trend estimates from these alternative models?

Figure 1.19 in Shumway (1988) shows a graph of the sulphur dioxide series. Compare that graph with the graph of sulphur dioxide in Figure 7.22. Notice that the peak value in the latter is less than 1500, whereas Shumway shows values in excess of 1800. Research Shumway *et al.* (1983) to determine if the figures we have taken from Shumway (1988) are correct.

The analysis reported in Shumway (1988) uses the logarithmically transformed sulphur series as an explanatory variable for excess deaths. It is stated in that text that the natural logarithm transformation is used, but if you examine the scale in Figure 1.20 there you will note that decimal logarithms were actually used. What is the effect of using one or other of the log transformations?

One tentative model determined by Shumway (1988, Chapter 4) relates excess deaths to current sulphur dioxide levels and to current and past temperatures,

$$y_t = 22.30 x_{1,t} + 2.10 x_{2,t} - 1.55 x_{2,t-2} + \nu_t,$$

where y is excess deaths, x_1 is log transformed sulphur dioxide level, and x_2 is temperature. What is wrong with this model? [**Hint:** consider changing the temperature measurement scale from Celsius as used here to Fahrenheit.] Compare alternative regression discount factors for the 'corrected' model. Is a static model reasonable? Examine the residuals from the one step ahead forecasts and the retrospective fit. Is there evidence of non-normality or serial correlation? How much of the variation in deaths is explained by the model?

Is there any evidence that including the one day lagged temperature variable would improve the forecasting power of the model?

It is suggested that temperature and pollution levels (the latter represented here by sulphur dioxide) have discontinuous effects on deaths. Partition the temperature and sulphur series into a series of bands, temperature in 3 degree bands for example. Examine deaths corresponding to the different bands. Can you detect any evidence for the hypothesis of discontinuous effect?

It is also suggested that changes in temperature might be an important contributory factor to deaths. Examine this hypothesis.

The data set LONDON.DAT includes two further variables, relative humidity and smoke level. Determine if these variables provide additional explanatory performance over what is achieved with the model above. Are these variables superior to sulphur and temperature?

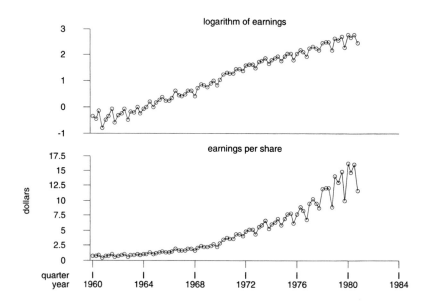

Figure 7.23 Quarterly earnings and logarithm of earnings from Johnson & Johnson.

7.12 Share Earnings

Figure 7.23 shows the quarterly share earnings (together with the logarithmic transformation) from Johnson & Johnson from the first quarter of 1960 to the fourth quarter of 1981 (Shumway, 1988). The exponential growth in earnings is plainly apparent from the graph, as is the increasing amplitude of the seasonal pattern. Actually, the seasonal pattern is more complicated than simply increasing in amplitude with time, as can be seen from the logarithm series. The seasonal pattern is pronounced in the early years but, unusually, decreasing in amplitude up to the early 1970s. Thereafter the seasonal amplitude begins to explode.

The estimated underlying trend in log earnings is shown in Figure 7.24 (the model is a second order trend and full seasonal component DLM). On this log scale the trend is very nearly piecewise linear, comprising two segments with the break in 1972. The rate of increase declining from 0.045 (annual earnings growth rate of 4.6%) in the pre-1972 period to 0.038 (3.9%) in the post-1972 period. The break coincides with the change in seasonal amplitude from a steadily decreasing pattern to a steadily increasing pattern. This can be seen from Figure 7.25. It is very clear that the period around the early 1970s was something of a watershed for the company.

7.12 Share Earnings

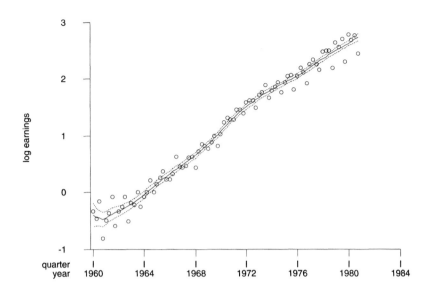

Figure 7.24 Estimated underlying trend in log earnings.

The lower panel in Figure 7.25 gives an excellent view of the dynamic development of the seasonal pattern in log earnings from 1960 to 1981. The amplitude changes already described are self-evident, but there are other changes too. Until 1965 the seasonal pattern was flat in the first half of the year, strongly positive in the third quarter, and offsettingly negative in the fourth quarter. Beginning in 1965 the seasonal downswing begins to be spread across the winter months (quarters one and four). By 1966 the first quarter has become more negative than quarter four. This development goes into reverse in 1968, a year that also sees the second quarter beginning to contribute a noticeable positive factor. In 1975 the effect of the second quarter is stronger than the third; but just three years later the first quarter becomes the largest positive factor. From 1972 onward only the fourth quarter is seasonally depressed.

Analysis Suggestions

Does it seem to you that the estimated level in the latter years is too high? What is the feature of the data that leads to this impression? Do you believe that the estimated level is too high?

The descriptive analysis detailed above utilised a DLM comprised of a linear growth trend component and a free-form seasonal component. The seasonal component is highly dynamic as we have seen, the trend really

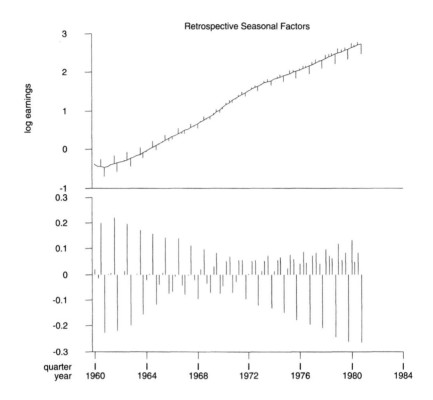

Figure 7.25 Estimated seasonal pattern in log earnings.

only changing in rate of growth around 1972. Such an analysis is fine for extracting estimates of the components, but how useful is it for prediction? It is a reasonably safe bet that extrapolating the underlying level into 1982 would be uncontroversial. But what about the cyclical variation about the trend? Unmodified, the model will simply superimpose the seasonal pattern estimated at the end of 1980 on to the extrapolated trend. Historically this is not a reasonable procedure. Without further modelling, suggest a suitable modification of the seasonal pattern. How well would your modifications have done if applied at the end of 1980 to forecast log earnings in 1981?

Consider modelling the seasonal amplitude development since 1970. The typical approach to handling a seasonal amplitude increasing with level is to make a transformation of the data: just as we have already done from

7.12 Share Earnings

the raw earnings series. What effect would such a transformation have on the currently (approximately) linear trend?

An alternative to transformation is the multiplicative trend-seasonal model mentioned in the air freight volume example above. Use the program you developed there to investigate the suitability of the model for the log earnings series (from 1972 on). Make forecasts for 1981 from the standpoint of the end of 1980. How do your forecasts from this model compare with your adjusted forecasts from the linear additive model above?

Shumway (1988) estimates a nonlinear model for the raw earnings series with additive trend and seasonal components, where the trend has an exponential growth form

$$\text{level}_t = \phi \text{level}_{t-1} + \omega_t$$

(and ω_t is an unknown stochastic term). It is the unknown multiplicative growth factor ϕ that makes this model nonlinear. The model may be linearised by using an estimate of the growth rate at each time. For forecasting at time t we could set this estimate to be

$$\phi_{t+1} = \frac{E[\text{level}_t | \text{Data}_t]}{E[\text{level}_{t-1} | \text{Data}_t]}.$$

Operationally we might simplify this estimate by using the on-line estimated level at time $t-1$, $E[\text{level}_{t-1}|\text{Data}_{t-1}]$, in the denominator rather than the one step back filtered estimate.

Write a computer program, or modify BATS, to implement the exponential growth model. Compare it to a DLM with a quadratic growth component for the earnings series. Compare also the exponential trend multiplicative seasonal model.

Suggest a model that might be used for the 1960–71 portion of the earnings series. Implement this model by suitably modifying your exponential growth multiplicative seasonal program. How good is your model at forecasting the earnings series?

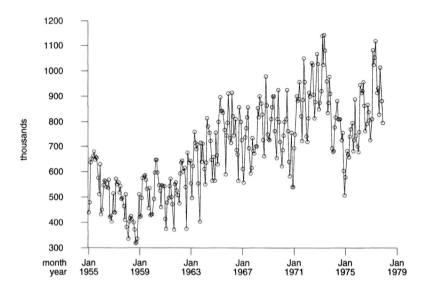

Figure 7.26 Monthly U.S. retail sales of passenger cars.

7.13 Passenger Car Sales

Figure 7.26 shows the monthly retail sales of passenger cars in the U.S. from January 1955 to December 1977 (Abraham and Ledolter, 1983, p. 398). Figure 7.27 shows the quarterly earnings for the world's largest automobile manufacturer, General Motors. Consider the sales series on the quarterly scale. How strong is the relationship between GM earnings and retail sales? Are the structural shifts in the sales series reflected in the GM earnings series?

Figure 7.41 shows the Standard and Poor's 500 stock price index, and index of U.S. industrial production for a period including the period of the data here. Is there a strong correlation between either of the indices and either the sales or earnings series? Explain. Estimate the trends in the sales and earnings series. Compare them with the index series.

How regular are the cyclical patterns in the sales and earnings series?

All of the series show a marked reduction in level occurred somewhere around 1974–75. Estimate the timing and magnitudes of the changes. What major economic changes happened in the early 1970s that would explain the level decreases in these series?

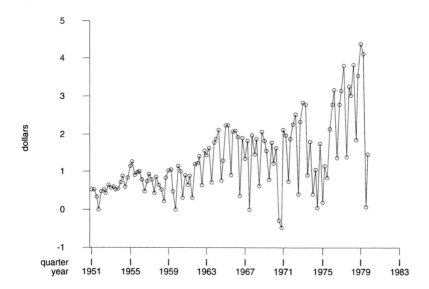

Figure 7.27 Quarterly earnings per share (stock split and dividend adjusted) for General Motors stock.

7.14 Phytoplankton

Figure 7.28 shows four series of four-weekly averages made at Lake Mendota (Lathrop and Carpenter, 1992): total phosphorous, stability index, *Daphnia* (water flea) biomass, and blue-green algae. (Note that the data here are calendar monthly averages; Lathrop and Carpenter work with four-week averages giving 13 'tetraweeks' per year.) There is substantial interest in modelling and forecasting the dynamics of blue-green algae contamination because the algae is a serious water quality nuisance. Stability is the amount of work the wind must do on the surface of the lake to mix the water thoroughly. It provides a physical index of seasonality because the lake is most stable in the summer and least stable in the winter. *Daphnia* is an important grazer that can reduce the amount of blue-green algae under some circumstances. Total phosphorous is the most important nutrient for the algae. It builds up during the winter, six to eight months before the worst algae blooms of the summer.

Analysis Suggestions
Build a regression model for the blue-green algae series including the stability index, *Daphnia* biomass, and (lagged) phosphorous series as explana-

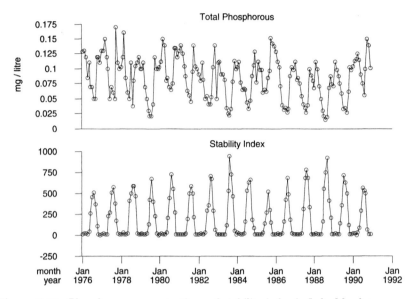

Figure 7.28a Phosphorous concentration and stability index in Lake Mendota.

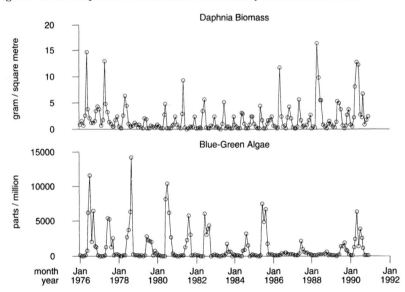

Figure 7.28b *Daphnia* biomass and blue-green algae concentration in Lake Mendota (Lathrop and Carpenter, 1992).

7.14 Phytoplankton

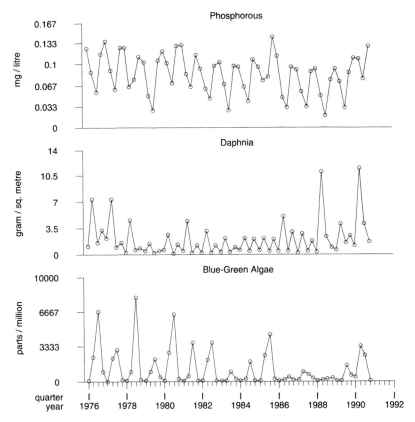

Figure 7.29 Quarterly blue-green algae, *Daphnia*, and phosphorous concentration.

tory variables. Compare results with the alternative model that excludes the stability index and uses a form seasonal component instead. Is there evidence of seasonality in addition to the stability component?

It has been suggested that a more suitable modelling approach is to use the logarithm transformed algae series rather than the raw series. Investigate.

Figure 7.29 shows the quarterly averaged blue-green algae, *Daphnia*, and phosphorous series. Examine models for the the quarterly series. Is there more identifiable structure on this time scale? Consider also averaging on a six month scale, beginning the aggregation in April to obtain a spring/summer and autumn/winter series.

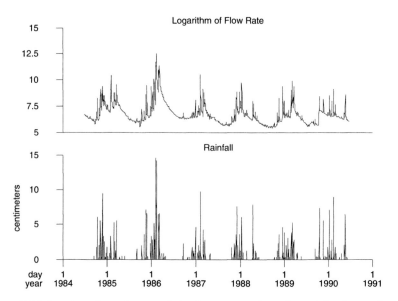

Figure 7.30 Log transformed mean daily flow rate and total precipitation at San Lorenzo Park (Rojstaczer and Wolf, 1991).

7.15 Further Data Sets

In the pages that follow there are time plots of several more of the data sets that are included on the BATS diskette. The plots illustrate the wide range of behaviour that is observed in time series, and the breadth of subject areas where time series data occurs. Of course, the subject area coverage is by no means exhaustive!

The sources from which the data sets have been culled are listed in the final section, each figure legend referencing the relevant source. Enjoy!

7.15 Further Data Sets

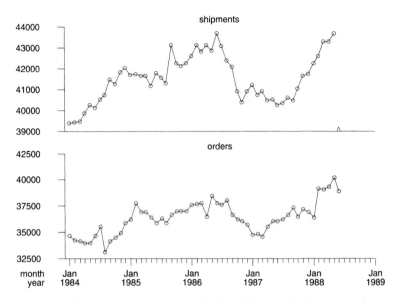

Figure 7.31 Monthly orders and shipments of valves. (Source: Pankratz, 1991.)

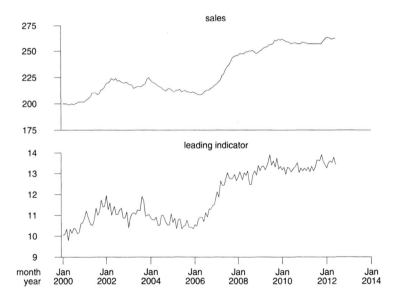

Figure 7.32 Sales data with leading indicator. (Source: Box and Jenkins, 1976; from Pankratz, 1991.)

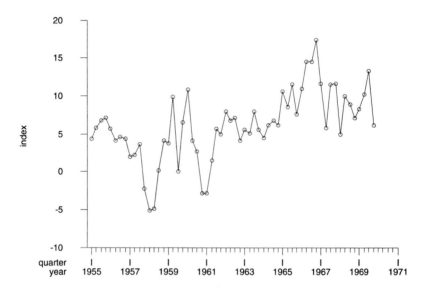

Figure 7.33 Quarterly change in business inventories. (Source: Pankratz, 1991.)

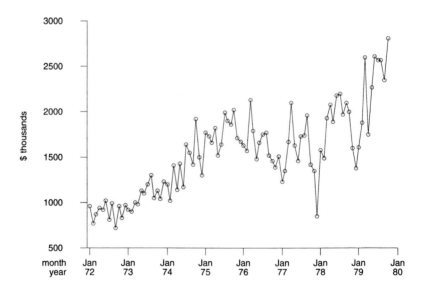

Figure 7.34 Monthly demand for repair parts for a large heavy equipment manufacturer in Iowa. (Source: Pankratz, 1991; Abraham and Ledolter, 1983.)

7.15 Further Data Sets 205

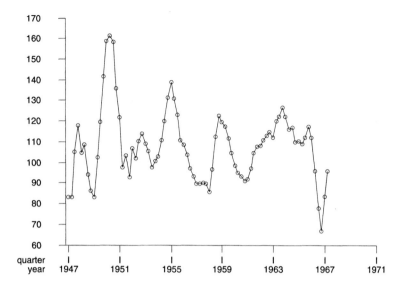

Figure 7.35 Quarterly, seasonally adjusted, housing permits in the United States. (Source: Pankratz, 1991.)

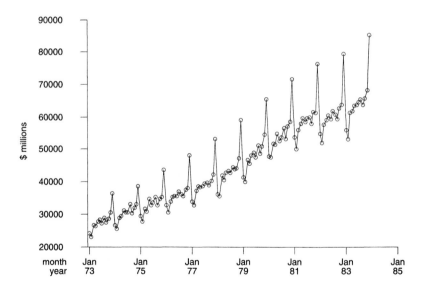

Figure 7.36 Monthly retail sales of nondurable goods stores. (Source: U.S. Commerce Department, Survey of Current Business; from Pankratz, 1991.)

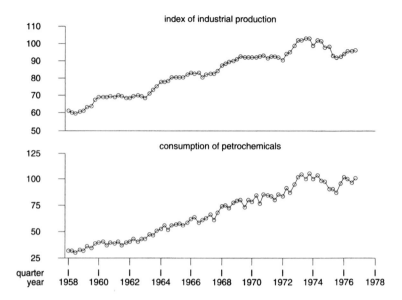

Figure 7.37 Quarterly U.K. consumption of petrochemicals, and seasonally adjusted index of industrial production. (Source: Pankratz, 1991.)

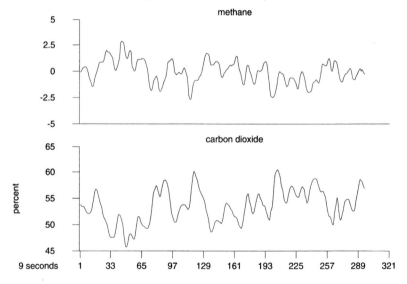

Figure 7.38 Methane gas input and concentration of carbon dioxide in gas furnace output. (Source: Box and Jenkins, 1976; from Pankratz, 1991.)

7.15 Further Data Sets

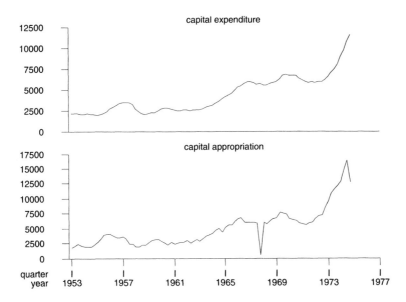

Figure 7.39 Quarterly capital appropriations and expenditures. (Source: Judge *et al.* 1982; from Pankratz, 1991.)

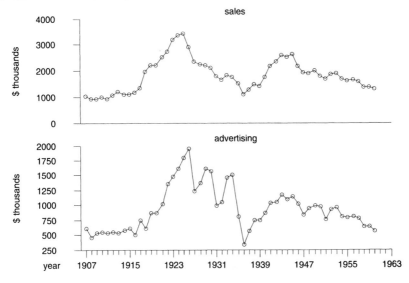

Figure 7.40 Annual sales and advertising expenditures for Lydia Pinkham. (Source: Vandaele, 1983; from Pankratz, 1991.)

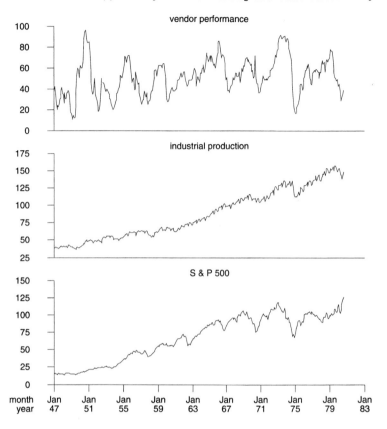

Figure 7.41 Monthly index of industrial production, Standard and Poor's 500 Stock Price Index, and Index of Vendor Performance. (Sources: U.S. Commerce Department, Survey of Current Business, and Business Conditions Digest; from Pankratz, 1991.)

7.15 Further Data Sets   209

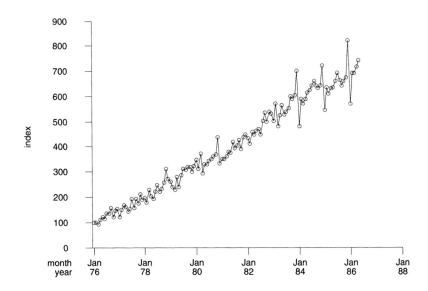

Figure 7.42 Monthly index of shipments of consumer products. (Source: Pankratz, 1991.)

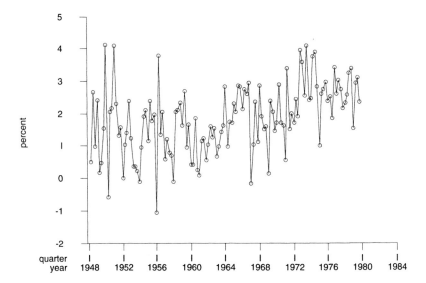

Figure 7.43 Quarterly growth rates of Iowa nonfarm income. (Source: Abraham and Ledolter, 1983.)

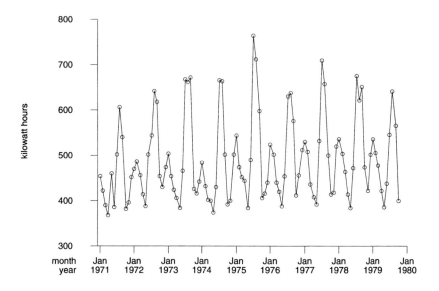

Figure 7.44 Monthly average residential electricity usage in Iowa City. (Source: Abraham and Ledolter, 1983.)

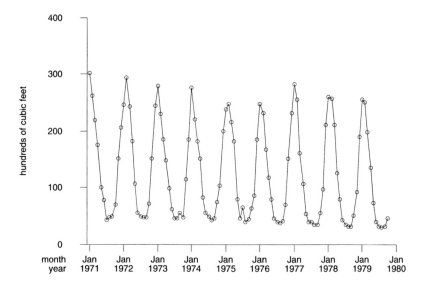

Figure 7.45 Monthly average residential gas usage in Iowa City. (Source: Abraham and Ledolter, 1983.)

7.15 Further Data Sets   211

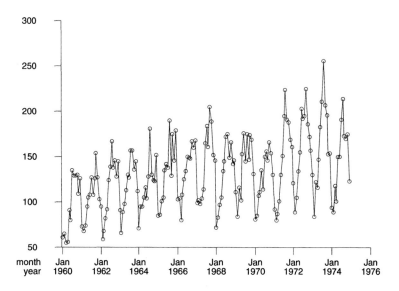

Figure 7.46 Monthly traffic fatalities in Ontario. (Source: Abraham and Ledolter, 1983.)

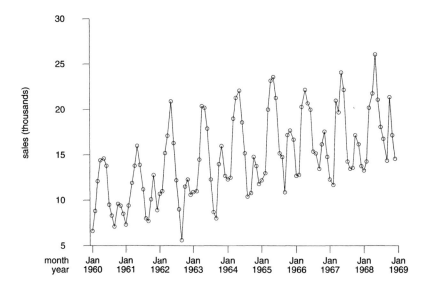

Figure 7.47 Monthly car sales in Quebec. (Source: Abraham and Ledolter, 1983.)

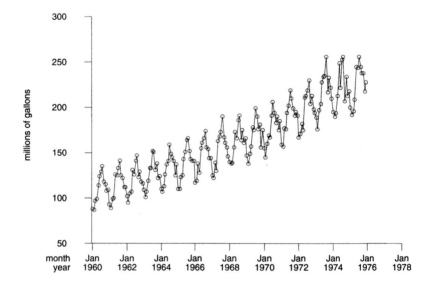

Figure 7.48 Monthly gasoline demand in Ontario. (Source: Abraham and Ledolter, 1983.)

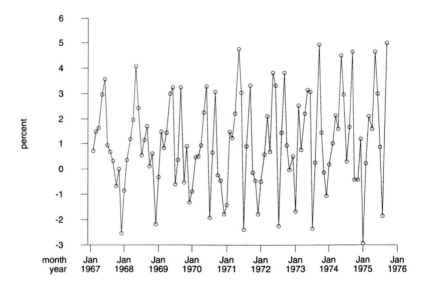

Figure 7.49 Monthly percentage changes in Canadian wages and salaries. (Source: Abraham and Ledolter, 1983.)

7.15 Further Data Sets   213

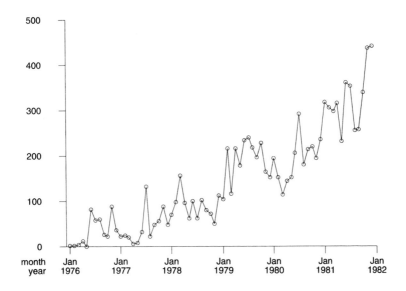

Figure 7.50 Monthly sales of computer software packages for undergraduate college-level curriculum development. (Source: Abraham and Ledolter, 1983.)

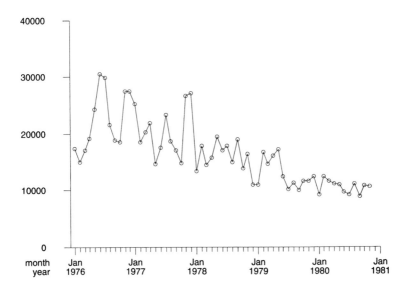

Figure 7.51 Monthly approvals of Master Card applications. (Source: Abraham and Ledolter, 1983.)

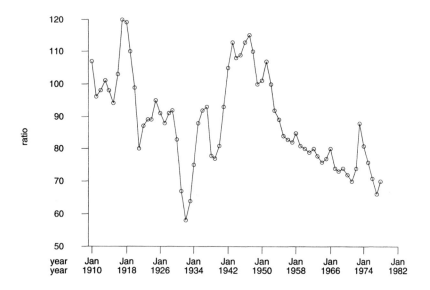

Figure 7.52 Annual farm parity ratio. (Sources: Historical Statistics of the United States and Statistical Abstract of the United States, U.S. Department of Commerce; from Abraham and Ledolter, 1983.)

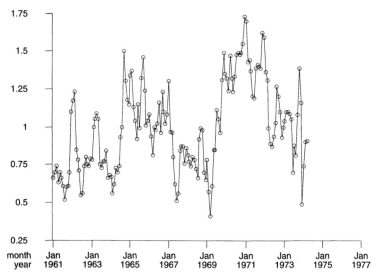

Figure 7.53 Monthly differences between the yield on mortgages and the yield on government loans in the Netherlands. (Source: Abraham and Ledolter, 1983.)

7.15 Further Data Sets 215

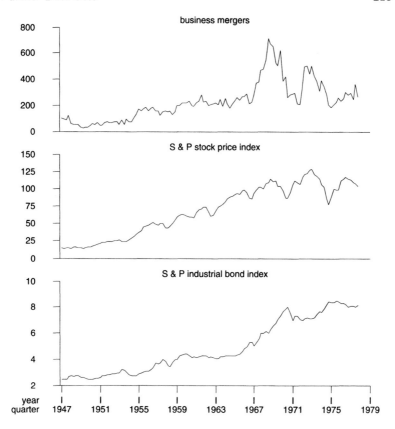

Figure 7.54 Quarterly business mergers, averages of monthly Standard & Poor's Stock Price and Industrial Bond indices. (Source: Abraham and Ledolter, 1983.)

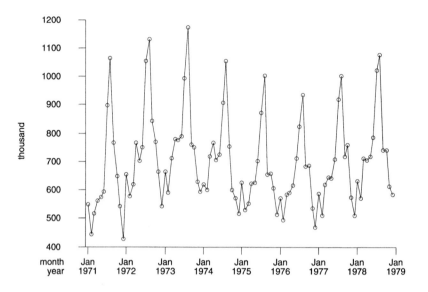

Figure 7.55 Monthly arrivals of U.S. citizens from foreign travel. (Source: Abraham and Ledolter, 1983.)

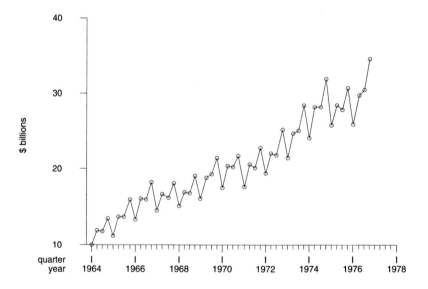

Figure 7.56 Quarterly plant and equipment expenditures in U.S. industries. (Source: Abraham and Ledolter, 1983.)

7.15 Further Data Sets 217

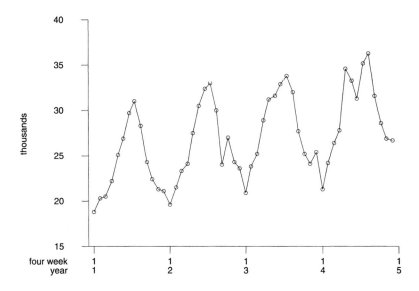

Figure 7.57 Four week totals of beer shipments. (Source: Abraham and Ledolter, 1983.)

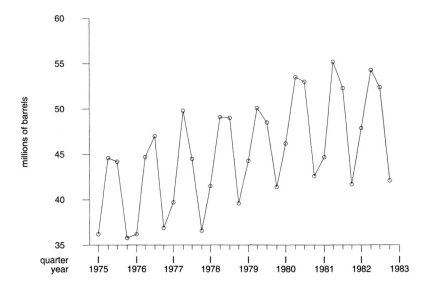

Figure 7.58 Quarterly U.S. beer production. (Source: Wei, 1990.)

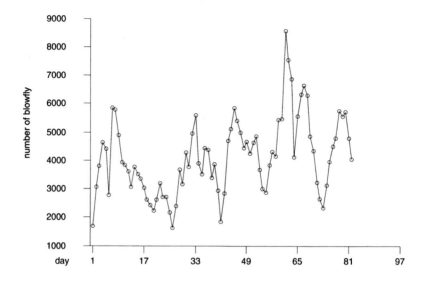

Figure 7.59 Blowfly data. (Source: Nicholson, 1950; from Wei, 1990.)

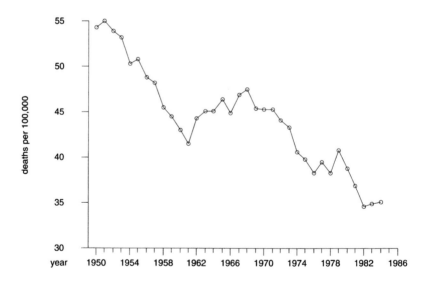

Figure 7.60 Yearly accidental death rate for Pennsylvania. (Source: Wei, 1990.)

7.15 Further Data Sets

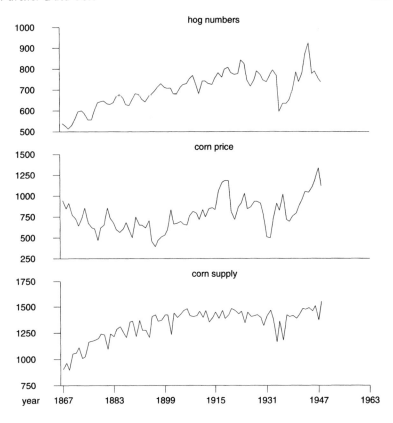

Figure 7.61 Yearly hog numbers, corn price, and corn supply. (Source: Wei, 1990.)

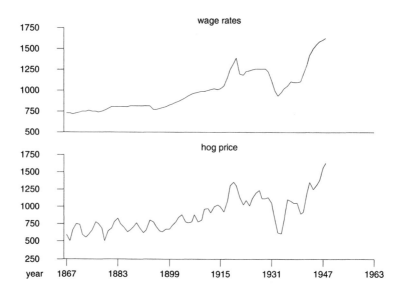

Figure 7.62 Yearly wage rates and hog prices. (Source: Wei, 1990.)

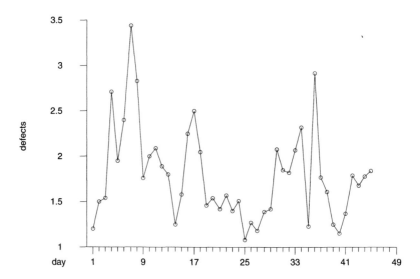

Figure 7.63 Daily average number of truck manufacturing defects. (Source: Wei, 1990.)

7.15 Further Data Sets

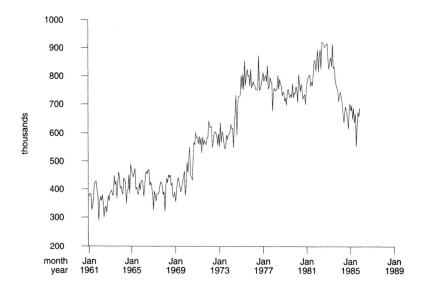

Figure 7.64 Monthly unemployed females between ages 16 and 19 in the U.S. (Source: Wei, 1990.)

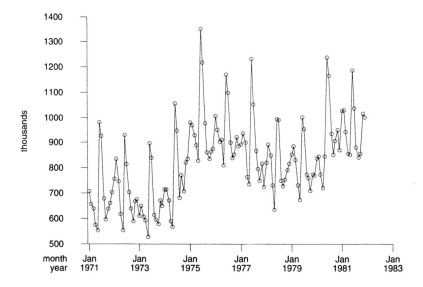

Figure 7.65 Monthly employed males between ages 16 and 19 in the U.S. (Source: Wei, 1990.)

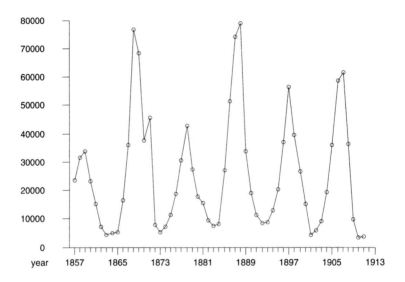

Figure 7.66 Yearly number of lynx pelts sold by the Hudson's Bay Company in Canada. (Source: Wei, 1990.)

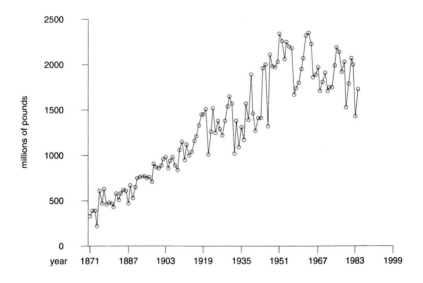

Figure 7.67 Yearly U.S. tobacco production. (Source: Wei, 1990.)

7.15 Further Data Sets

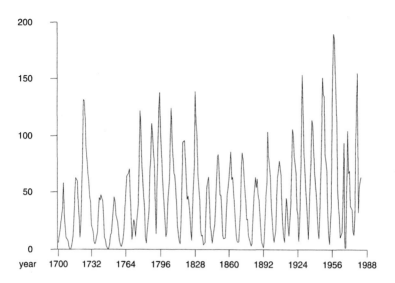

Figure 7.68 Wolf yearly sunspot numbers. (Source: Wei, 1990.)

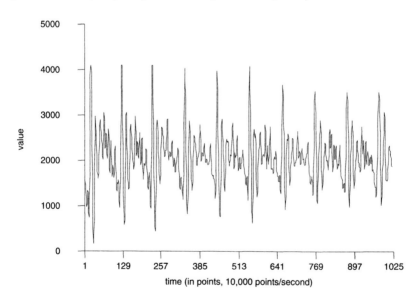

Figure 7.69 Speech record of the syllable *ahh* sampled at 10,000 points per second. (Source: Shumway, 1988.)

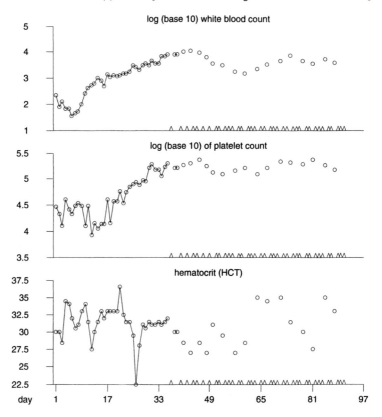

Figure 7.70 Bone marrow transplant data: white blood count, platelet, and hematocrit. (Source: Shumway, 1988.)

7.15 Further Data Sets   225

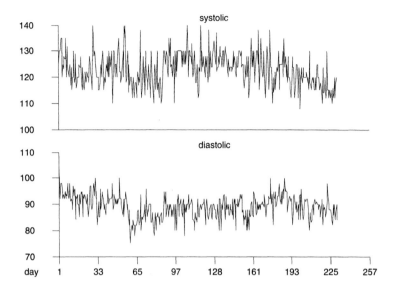

Figure 7.71 Twice daily blood pressure readings for a mild hypertensive. (Source: Shumway, 1988.)

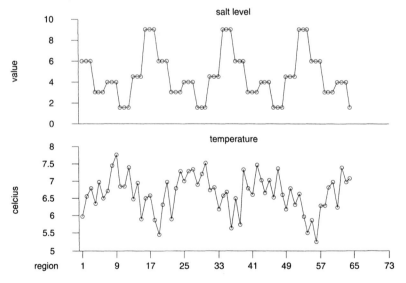

Figure 7.72 Spatially distributed soil samples: temperature reading and salt treatment level. (Source: Bazza, 1985; from Shumway, 1988.)

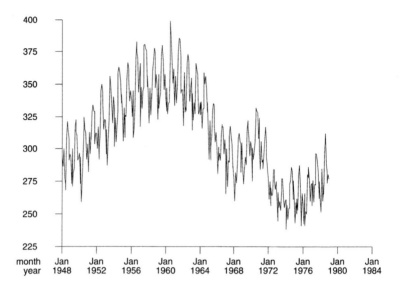

Figure 7.73 Monthly registered live births in the U.S. (Source: Monthly Vital Statistics Reporter, National Center for Health Statistics, U.S. Public Health Service; from Shumway, 1988.)

7.16 Data Set List

The BATS diskette contains all of the data sets from the following sources:

- Abraham and Ledolter, *Statistical Methods for Forecasting*, courtesy of John Wiley & Sons, Inc.
- Pankratz, *Forecasting with Dynamic Regression Models*, courtesy of John Wiley & Sons, Inc.
- Shumway, *Applied Statistical Time Series Analysis*, courtesy of Prentice-Hall.
- Wei, *Time Series Analysis*, courtesy of Addison-Wesley Publishing Company, Inc.

The following sections list the data sets by name, cross-referenced to the sources and the figures in this chapter.

Abraham and Ledolter

BEER.DAT Four weekly totals of beer shipments (thousands of units), series 8; Figure 7.57.
CARS_US.DAT Monthly United States retail sales of passenger cars (thousands), p. 398; Figure 7.26
CARSALES.DAT Monthly car sales in Quebec; series 4; Figure 7.47.
COMPUTE.DAT Monthly sales of computer software packages for undergraduate college-level curriculum development; p. 396; Figure 7.50.
ELECTRIC.DAT Monthly average residential electricity usage in Iowa City; series 3; Figure 7.45.
FARM.DAT Quarterly Iowa nonfarm income; series 1 and 2; Figure 7.43.
GAS.DAT Monthly average residential gas usage in Iowa City; series 11; Figure 7.45.
GASOLINE.DAT Monthly gasoline demand in Ontario; series 12; Figure 7.48.
GM_EARN.DAT Quarterly earnings per share (stock split and dividend adjusted) for General Motors stock; pp.410–411; Figure 7.27.
MCARD.DAT Monthly approvals of Master Card applications; p. 406; Figure 7.51.
MERGER.DAT Quarterly business mergers, averages of monthly Standard & Poor's stock price and industrial bond indices; pp.408–408; Figure 7.54.
PARITY.DAT Annual farm parity ratio; p. 406; Figure 7.52.
PLANT.DAT Quarterly plant and equipment expenditures in U.S. industries; series 7; Figure 7.56.
REPAIR.DAT Monthly demand for repair parts for a large heavy equipment manufacturer in Iowa; series 10; Figure 7.34.
TRAFFIC.DAT Monthly traffic fatalities in Ontario; series 5; Figure 7.46.

VISIT.DAT Monthly arrivals of U.S. citizens from foreign travel; p. 411; Figure 7.55.
WAGES.DAT Monthly percentage changes in Canadian wages and salaries; series 13; Figure 7.49.
YIELD.DAT Monthly differences between the yield on mortgages and the yield on government loans in the Netherlands; series 9; Figure 7.53.

Pankratz

CAPITAL.DAT Quarterly capital appropriations and expenditures; series 19 and 20; Figure 7.39.
CONSUME.DAT Monthly index of shipments of consumer products; series 31; Figure 7.42.
FREIGHT.DAT Monthly air carrier freight volume; series 30; Figure 7.15.
FURNACE.DAT Methane gas input and concentration of carbon dioxide in gas furnace output; series 17 and 18; Figure 7.38.
HOUSPERM.DAT Quarterly housing permits in the United States (seasonally adjusted); series 9; Figure 7.35.
INDPROD.DAT Monthly index of industrial production, Standard and Poor's 500 Stock Price Index, and Index of Vendor Performance; series 25, 26, and 27; Figure 7.41.
INVENT.DAT Quarterly change in business inventories; series 6; Figure 7.33.
KWHOUR.DAT Monthly index of kilowatt hours used, heating degree days, and cooling degree days; series 12, 13, and 14.
MTHHOUS.DAT Monthly United States housing starts and sales of single-family structures; series 23 and 24; Figure 7.10.
PETRCHEM.DAT Quarterly U.K. consumption of petrochemicals, and seasonally adjusted index of industrial production; series 15 and 16; Figure 7.37.
PINKHAM.DAT Annual sales revenue and advertising outlay for the Lydia Pinkham company; series 21 and 22; Figure 7.40.
QHSTART.DAT Quarterly housing starts; series 11.
RECEIPT.DAT Monthly U.S. Federal Government net receipts; series 7.
RETSALES.DAT Monthly retail sales of nondurable goods stores; series 10; Figure 7.36.
SALES.DAT Sales data with leading indicator; series 4 and 5; Figure 7.32.
SAVRATE.DAT Quarterly U.S. saving rate; series 3; Figure 7.13.
SCHIZO.DAT Daily perceptual speed score for a schizophrenic patient; series 29; Figure 7.12.
TELEFONE.DAT Monthly average daily calls to Cincinnati directory assistance; series 28; Figure 7.11.
VALVE.DAT Monthly orders and shipments of valves; series 1 and 2; Figure 7.31.

7.16 Data Set List

Shumway
BIRTH.DAT Monthly registered live births in the U.S.; Table 10; Figure 7.73.
BLOOD.DAT Twice daily blood pressure readings for a mild hypertensive; Table 6; Figure 7.71.
IND.DAT Monthly Federal Reserve Board production index and number unemployed (10,000s); Table 5.
JOHNSON.DAT Quarterly earnings and logarithm of earnings from Johnson & Johnson; Table 7; Figure 7.23.
LONDON.DAT Daily London mortality (adjusted), sulphur dioxide, and temperature series; Table 1; Figure 7.22.
MARROW.DAT Bone marrow transplant data: white blood count, platelet, and hematocrit; Table 9; Figure 7.70.
SOIL.DAT Spatially distributed soil samples: temperature reading and salt treatment level; Table 4; Figure 7.72.
SPEECH.DAT Speech record of the syllable *ahh* sampled at 10,000 points per second; Table 2; Figure 7.69.

Wei
BEER.DAT Quarterly U.S. beer production; series W10; Figure 7.58.
BLOWFLY.DAT Blowfly data; series W3; Figure 7.59.
DEATH.DAT Yearly accidental death rate for Pennsylvania; series W5; Figure 7.60.
DEFECT.DAT Daily average number of truck manufacturing defects; series W1; Figure 7.63.
EMPLOY.DAT Monthly employed males between ages 16 and 19 in the U.S.; series W9; Figure 7.65.
HOGS.DAT Yearly hog numbers, corn price, and corn supply; series W13; Figure 7.61.
LYNX.DAT Yearly number of lynx pelts sold by the Hudson's Bay Company in Canada; series W7; Figure 7.66.
SUNSPOT.DAT Wolf yearly sunspot numbers; series W2; Figure 7.68.
TOBACCO.DAT Yearly U.S. tobacco production; series W6; Figure 7.67.
UNEMP.DAT Monthly unemployed females between ages 16 and 19 in the U.S.; series W4; Figure 7.64.

Other Data Sets
CANDY.DAT Monthly sales and price of a confectionary product; Chapter 10.
CHEMICAL.DAT Monthly sales of a chemical product; Figure 11.21.
GAS.DAT Monthly inland natural gas consumption in the U.K.; Figure 7.5.
GREEK1.DAT Greek marriage series; Chapter 6.

GREEK2.DAT More Greek marriage series; Chapter 6.
MARRYUK.DAT Quarterly marriages in the United Kingdom; Figure 7.8.
NILE.DAT Annual volume of the Nile river; Figure 7.1.
PLANKTON.DAT Monthly water quality data; Figure 7.28–7.29.
QACCIDS.DAT Accident injury series; Chapter 11.
RPI.DAT United Kingdom retail price index; Figure 7.7.
SHARE.DAT Market share series; Chapter 5.
TOBACCO.DAT Monthly tobacco sales; Figure 7.6.
TURKEY.DAT Monthly sales of turkey chicks in Eire; Chapter 4.
WATER.DAT Daily rainfall and river flow; Figure 7.30.

7.17 References

Abraham, B. and Ledolter, J. (1983). *Statistical Methods for Forecasting.* Wiley, New York.

Box, G.E.P. and Jenkins, G.M. (1976). *Time Series Analysis: forecasting and control.* Holden-Day, San Francisco.

Balke, N.S. (1993). Detecting level shifts in time series, *Journal of Business & Economic Statistics*, **11**, 1, 81–92.

Bazza, M. (1985). *Modelling of soil surface temperature by the method of two dimensional spectral analysis*, Ph.D dissertation, University of California, Davis.

Carlstein, E. (1988). Nonparametric change-point estimation, *The Annals of Statistics*, **16**, 188–197.

Cobb, G.W. (1978). The problem of the Nile: conditional solution to a change-point problem, *Biometrika*, **65**, 243–251.

Dümbgen, L. (1991). The asymptotic behaviour of some nonparametric change-point estimators, *The Annals of Statistics*, **19**, 1471–1495.

Harrison, P.J. (1988). Bayesian forecasting in O.R. In *Developments in Operational Research*, eds. N.B. Cook and A.M. Johnson. Pergamon Press, Oxford.

Harrison, P.J. and Stevens, C.F. (1976). Bayes forecasting in action, Warwick Research Report 14, Department of Statistics, University of Warwick, Coventry, England.

Judge, G.G.R., Hill, R.C., Griffiths, W.E., Lütkepohl, H., and Lee, T.C. (1982). *Introduction to the Theory and Practice of Econometrics.* Wiley, New York.

Lathrop, R.C. and Carpenter, S.R. (1992). Phytoplankton and their relationship to nutrients. In *Food Web Management: A Case Study of Lake Mendota*, ed. J.F. Kitchell. Springer-Verlag, New York.

7.17 References

Nicholson, A.J. (1950). Population oscillations caused by competition for food, *Nature*, **31**, 476–477.

Pankratz, A. (1991). *Forecasting with Dynamic Regression Models*. Wiley, New York.

Rojstaczer, S., and Wolf, S. (1991). Hydrologic Changes Associated with the Loma Prieta Earthquake in the San Lorenzo and Pescadero Drainage Basins, U.S. Department of the Interior U.S. Geological Survey, Open-File Report 91-567.

Shumway, R.H. (1988). *Applied Statistical Time Series Analysis*, Prentice-Hall, Englewood Cliffs, NJ.

Vandaele, W. (1983). *Applied Time Series and Box-Jenkins Models*, Academic Press, New York.

Wei, W.W.S. (1990). *Time Series Analysis*. Addison-Wesley, Reading, MA.

West, M. and Harrison, P.J. (1989). *Bayesian Forecasting and Dynamic Models*, Springer-Verlag, New York.

Part B

**INTERACTIVE
TIME SERIES ANALYSIS
AND FORECASTING**

Chapter 8

Installing BATS

This chapter details the information you need in order to run the BATS program and how to use the documentation in Parts B and C of the book. Please read the material carefully: it is designed to help you avoid problems and when problems do occur it will indicate where to look for a solution.

8.1 Documentation

The documentation is in two parts: this first part is a user guide, and the second part, Part C of the book, is a reference guide.

8.1.1 User Guide

Part B of the book contains descriptions of the equipment you need in order to be able to use the BATS program; how to configure your computer for use with the program; and a series of tutorial examples to guide you through the stages of available analyses. As we stated in the Preface, the tutorials presented here are designed to be read in parallel with the material in Part A. However, the parts are self-contained so that if you are already familiar with the DLM, or just want to go ahead and see what the program and the techniques it embodies can do, then these tutorials can be worked through without reading Part A.

8.1.2 Reference Guide

The reference guide in Part C provides a complete listing of all of the facilities in the program with a description of what each facility provides and how it works. The material there will really only be useful to you once you have become familiar with the basic program operation, by working through at least some of the tutorials here in Part B for example.

8.2 How to Use This Guide

Whatever your knowledge of computers and time series analysis you should read this chapter. It contains the information you need to load and run BATS on your computer and how to operate the program effectively with your machine's particular configuration. Chapter 9 provides your introduction to using BATS. You will need to read this material—and follow the instructions—to learn how to communicate your requirements to the program.

How you use Chapters 10 and 11 depends upon your knowledge of Bayesian analysis and the dynamic linear model. Chapter 10 is structured as a guided tour of the basic modelling facilities of the univariate DLM; how to tell BATS which facilities you want to include in your model; and how to perform time series analysis and forecasting with your model. The tour shows you the way, and along the way provides (we hope) illuminating explanation. If you are already familiar with the principles of Bayesian analysis and dynamic models you may wish to skip the discussion and just follow the summary lists of commands. (You will probably want to read at least the introductory remarks to each section of the tour so as to get your bearings and the intended heading before starting out.) If you are not too familiar with the practical application of Bayesian dynamic modelling, then you will benefit most from reading the tutorial discussion carefully as you work through the examples. Reviewing the material in Part A of the book after completing a tutorial will help reinforce concepts, model structure, and methodology.

Chapter 12 deals with the problem of missing data. If your data is incomplete or contains untrustworthy observations, then you will need to read the material here. The chapter explains how to tell the program which of your data values are 'missing' and how the program displays missing values for you.

Chapter 13 contains the material you will need when you want to begin analysing your own data. It describes the requirements for structuring your data files so that BATS will understand them, and how to read files produced by the program.

8.2 How to Use This Guide

The tutorials in Chapters 9 through 13 are designed to be 'stand alone' in the sense that each begins from the point of a fresh BATS session. However, the later tutorials do assume knowledge of the material in the preceding chapters and there are several references to 'what we have seen before'. So the best way to utilise the tutorials is to work through them in the order presented. Of course, once you have been through them all, you will be able to dip in at random to remind yourself of any particular feature.

The analyses detailed in the tutorials have been designed to illustrate the practice of dynamic modelling, to demonstrate the facilities that BATS provides to accomplish this, and of course to show you how to use these facilities. Although the text includes motivation for the analyses performed and discusses the results obtained, the tutorials are not intended to be definitive examples of time series analysis or forecasting. That is the proper domain of case studies and is best elucidated without the encumbrance of program usage instructions: see the applications in Part A of the book.

After reading through this guide and working through the tutorials you will be in a position to use BATS to do case study analyses of your own.

8.2.1 Typographical Conventions

Three different styles of printed type are used to distinguish material. The standard typeface that you are reading at the moment is used for general description, discussion, and advice. A *slanted* variation of this typeface is used to emphasise something to which we particularly want to draw your attention. And a `typewriter` like typeface is used for distinguishing program commands and user input from the general text.

Lists of points, remarks, or instructions are interspersed throughout the text. We have used another typographical device to 'set off' these lists and to distinguish between remarks and instructions.

- This is an example of an itemised remark.

An itemised remark is a point we wish to make, not an instruction to do something with the program. An instruction to do something with the program looks like this:

▷ This is a command list entry.

The two types of item are not mixed in any list in the guide.

8.3 Before You Begin

To use BATS you will need the following equipment.

- A computer which is IBM PC compatible (IBM PC, XT, AT, PS/2, and clones) or which may emulate the PC operating environment.
- DOS 3.0 or later.
- Adequate free disc storage (free space on a fixed disc drive, or formatted floppy discs with free space). BATS creates and uses disc files whose sizes vary with the data set length and model complexity. A fixed disc is recommended for storing these files and at least 2MB of free disc space should be available.
- A graphics adapter and suitable monitor. The supported graphics adaptors are IBM CGA, EGA, and VGA; Hercules monochrome; and ATT (Olivetti) monochrome. If you are using the Microsoft Windows version of the program you need not be concerned with graphics adaptors: Windows takes care of all that.

Analyses performed in BATS are quite computationally intensive and predominantly floating point. Therefore, it is advisable to have a math co-processor installed in your computer. BATS will operate correctly without such hardware floating point support, but performance, especially during the filter operations, will be noticeably poorer.

In the DOS version of the program the mouse (or pointing device) is supported in graphics mode but not in text mode. Although it is not necessary to have a mouse in order to utilise the full range of interactive graphics options, it is more convenient.

8.4 Installing BATS

The BATS diskette contains the following files:

- **BATS.EXE** This is the BATS program; it is an executable file similar to the .EXE files provided with the disc operating system (DOS).
- **BATS.PS1** This is a readable text file which contains PostScript procedure definitions which are required to produce PostScript output files for graphs.
- Over 50 data sets included on the diskette including all the series analysed in the book. See Chapter 7.
- **LATEST.TXT** Contains information that is not in the book. The most recent bug reports are included along with details of recent changes or additions to the program. You should read this file before using BATS to avoid any unnecessary difficulties.

If your computer contains a fixed disc drive you should copy all of the files on the BATS diskette to a suitable location on that drive. For example, if you want your BATS files to reside in a separate directory called BATS at the root of drive C (your fixed disc drive) follow these steps (you must press the ENTER or RETURN key after typing each command):

- At the DOS prompt type cd \ to place the system at the root directory of that drive.
- Type mkdir BATS to create a new directory at the root of the file system for the BATS files.
- Type cd BATS to move to the newly created directory.
- Making sure that your BATS diskette is placed in the appropriate floppy drive, say drive A, type copy a:*.* to copy the contents of the BATS diskette to the fixed disc directory.

If your computer does not have a fixed disc drive, then you should prepare three new diskettes as follows:

- Disc 1: copy BATS.EXE from the BATS diskette. This will be the diskette from which you will start a BATS session. You may also want to copy the example data files to this diskette.
- Disc 2: copy BATS.PS1 from the BATS diskette. This will be the diskette on which your PostScript output files will be created.
- Disc 3: empty. This will be the BATS session work diskette on which the analysis files ONLINE.RES and SMOOTH.RES will be generated. The storage capacity of this diskette will determine the limits of modelling ability (that is, data set size and model form) for your computer setup. The larger the diskette capacity the better.

Whatever the disc configuration of your computer system, you should keep the BATS diskette in a safe place. It will be needed if your working copies (on fixed or floppy disc) become damaged or corrupt.

8.5 Using BATS with Floppy Disc Drives

If your computer has only a single floppy disc drive and no fixed drive you will have to exercise some caution in order to get the most from BATS. To load the program, put the diskette containing BATS.EXE into the drive and type BATS (followed, as all DOS commands must be, by pressing the ENTER or RETURN key). Once BATS is running and the initial screen is displayed you may replace the program diskette in the disc drive with a data file diskette. After loading your chosen data set into BATS, remove the data file diskette and replace it with an empty diskette (see Disc 3 above). This 'work' diskette should only be removed when you wish to read or write a data file or create a PostScript output file. Chapter 9 describes the data file loading procedure; you may need to refer back to this material when you work through the tutorial.

If your computer is equipped with two floppy drives you can avoid much of the diskette-swapping necessary with a single drive configuration. One drive may be used for the program diskette, followed by the work diskette once BATS is running; and the other for data and PostScript files. You must be sure to configure BATS so that the appropriate drive is used for each of these file types, or you may end up with a data file diskette containing analysis results. This is not a problem, but during an analysis your diskette will very likely become full, causing the analysis to abort. The appendix to Chapter 9 describes how to instruct BATS about your directory and file structures.

8.6 Using BATS with A Fixed Disc Drive

BATS allows files to be accessed and created in any valid drive/directory combination so that it is not necessary either to start the program from its home directory (which we will assume is C:BATS for convenience), or to maintain data files in one location. There are no restrictions on the allowable file names or extensions (other than imposed by DOS); however, it is usually good practice (for the peace of mind of your co-workers as well as yourself!) to adhere to a consistent naming convention. The default file extensions which are set on the BATS diskette as distributed (but which may be reconfigured to suit your own needs) are .DAT and .FRE. In addition, PostScript output files are identified by the extension .PS. Filenames with the .DAT extension are used for data files written in the BATS format; that is, data files with data descriptive information included at the beginning. File names with the .FRE extension are used for data files that contain only columns of numbers and no other information. Part C provides a complete description of these file formats.

Directory structure will vary depending on your particular activities but common arrangements include:
- All data files together, for example in C:BATS\DATA; PostScript output files together, for example in C:BATS\PS.
- Directories arranged according to project, with data and PostScript files either collected together or separated as above.

It is not our purpose to suggest how you should organise your life, but we can pass on a message borne of extensive experience: following some strategy of file organisation will make your life much more trouble-free than otherwise. The appendix to Chapter 9 describes how to instruct BATS about your directory and file structures.

8.7 Using Expanded/Extended Memory

BATS is not able to make use of either expanded or extended memory. However, if your computer does have such memory installed, it is possible to increase the efficiency of the major operations by using a 'RAM-disc'. Using a RAM-disc for the filter results files, ONLINE.RES and SMOOTH.RES, will increase the speed with which the filter operations are performed. See your DOS manual for details of how to configure and use a RAM-disc. *Note that you should not use a RAM-disc for permanent file storage, that is, for data files or PostScript output files, since they will be lost when you reboot or power off your computer.* If you are a Windows user, ignore these comments and see the next section.

8.8 Microsoft Windows

There are two versions of the BATS program supplied on the disc accompanying the book: a vanilla DOS version, BATS.EXE, and a Microsoft Windows version, BATSWIN.EXE. If you attempt to use the Windows version from DOS you will get an error message.

If you are already a Windows user, then we need not advise you of the benefits of working in that environment. If you are not, we strongly recommend that you try it out: particularly with BATS!

The programs contain essentially the same functionality and menu structure, the minor differences being noted as they are encountered. There are also one or two differences in flavour. In the DOS version, only one menu is visible at any one time. It is necessary to explicitly enter key strokes to move up and down the menu tree; this is explained in Chapter 9. In the Windows version, the main menu is visible at all times with subordinate menus popping up as you select items on the parent. If you are already familiar with Windows applications, you will be aware of how things work.

Since space precludes giving explicit instructions for both program versions, in the following chapters we have taken the path of giving the instructions to follow when using the DOS program. If you work through Chapter 9 with the DOS program then switch to the Windows program and repeat the exercise (it takes only a few minutes), you will easily see how to interpret the tutorials directly in Windows. By the time the next edition of this book appears, DOS will probably have gone the way of the dinosaurs and the Windows version will be explicitly detailed.

Chapter 9

Tutorial: Introduction to BATS

Chapter 8 described the basic requirements and procedures for setting up your computer to run BATS. This chapter presents an overview of the way the program operates, how it receives instructions, and how it responds to them. You will learn how to drive BATS where you want it to go.

9.1 Starting a Session

If you have read Chapter 8 and installed BATS as described therein, all that is necessary to begin a session is:
- ▷ Make sure that the active directory is the one in which **BATS.EXE** has been placed (you may need to use the change directory command, **cd**, to set this).
- ▷ Type **BATS** and press the **ENTER** key at the **DOS** prompt.

The BATS startup display will appear on the screen and will remain showing until you enter a keystroke: any key will do. When you press a key the startup display is replaced with the basic BATS menu screen.
- ▷ Press a key to see the starting menu.

9.2 Operating with Menus

BATS menus consist of two 'bars' in the upper part of the screen (the 'control' area) the first of which displays space separated options while the second displays a brief explanation of the currently highlighted option.

244   Applied Bayesian Forecasting and Time Series Analysis

```
BATS 2.10              SELECT              Root menu
Available memory:
[Data] Configuration Exit
Data handling and examination
```

Figure 9.1 Startup menu.

Figure 9.1 shows the control portion of the screen as it appears when BATS is initially loaded (after the startup display is removed). Notice that there are just three options in the menu bar, `Data`, `Configuration`, and `Exit`, with the leftmost highlighted. The lower menu bar—option help bar—gives a description of this `Data` option. You can move through the menu options by using the left and right arrow cursor movement keys on your computer's keyboard.

  ▷ Press the right arrow key.

Notice that the option help text changes as you move through the different options.

  ▷ Press the right arrow key again to highlight `Exit`.
  ▷ Press the right arrow key once more.

Notice that the movement is 'wrap-around', that is, with the rightmost option highlighted pressing the right arrow key will make the leftmost option the current selection. Similarly, with the leftmost option highlighted pressing the left arrow key will make the the rightmost option the current selection.

Selecting an option may be done either by (i) first making it the current selection by using the cursor movement keys as just described, and then pressing the `ENTER` or `RETURN` key, or (ii) pressing the key corresponding to the first letter of the option name (case does not matter). Try this example:

  ▷ Use the cursor movement keys to select `Exit` as the current option.
  ▷ Press `ENTER`.

You are asked to confirm the request to terminate the BATS session. We do not want to leave BATS at this stage so

  ▷ Press `N` (or `n`) to select 'No'.

Figure 9.2 Root and first level of menu hierarchy on startup.

9.3 Traversing the Menu Hierarchy

The menu structure is arranged hierarchically. Starting from the initial menu—which we will refer to as the 'root' from now on—a series of levels each with its own set of submenus (or sublevels) is encountered. The only way to move between levels is to traverse the menu 'tree'. It is not possible to jump arbitrarily between levels to go from a given position to any other desired position.

Take a look at Figure 9.2 which displays part of the menu structure. The first two levels of the hierarchy are shown as they appear when BATS is first loaded.

▷ Select **Data** from the root menu.

(If you have been following this tutorial so far, the root menu should be displayed.) Notice that the **Data** menu now appears with the leftmost option highlighted as the current selection. The leftmost option is always highlighted on entering a new menu.

▷ Select **Quit** to return to the root.

You may examine each of the other level one menus in turn by selecting them from the root as we have just done for **Data**. In each case the **Quit** option will return control one level up the hierarchy—back to the root in this instance.

The hierarchical structure of the menus means that you may not directly select an option from one menu—say **Store** from the level one **Configuration** menu—while another menu is currently active—say the **Data** menu. Movement from menu to menu can only be done strictly in accordance with the hierarchy. To get from **Data** to **Store** you have to (i) **Quit** from **Data**, returning to the root; (ii) select **Configuration** from the root; (iii) then select **Store**.

Let's go ahead and have a look at some time series data. For this part of the tutorial you will need the example data files supplied with BATS.

You may wish to review the material on disc drive usage in Chapter 8 before continuing. If you are using a floppy disc drive system, place the diskette containing the example data files into the disc drive from which you loaded the program. Should you have a fixed disc drive and if the data files are in the same directory from which the program was loaded, you can go directly to the next Section. Otherwise you will have to exit from the program, move to the directory containing the data files, and restart. Exiting the program is straightforward.

▷ Quit successively to the root.
▷ Select **Exit**.
▷ Select **Yes**.

Control returns to the computer's operating system. At this point in your experience with BATS the easiest way to access the data sets is to put them in the same directory as the program file, then start BATS from there.

9.4 Examining Data

To view a data set it will first of all need to be loaded into the program. This means reading the file containing the data. Here's how. Beginning at the root make the following selections in turn.

▷ **Data**
▷ **Input-Output**
▷ **Read**
▷ **Bats-format**

The screen should now be displaying a list of all those files that have the .DAT extension which exist in the directory from which BATS was loaded. This list should include the example data sets, **CANDY.DAT**, **MARRIAGE.DAT**, and **QACCIDS.DAT**. If these files are not showing, check that you have correctly followed the instructions in the above list.

The list selection procedure is similar to the menu option selection procedure already described. The differences are (i) list items are displayed in columns so that vertical movement among items in a column is possible in addition to horizontal movement between columns; (ii) items *cannot* be selected by typing the first letter: the desired item must be highlighted by using the cursor movement keys then selected by pressing **ENTER** or **RETURN**. When a list is initially displayed, the first entry in the leftmost column is highlighted.

▷ Select **CANDY.DAT**. (If you only have the example data files with name extension .DAT, then **CANDY.DAT** is already highlighted so just press **ENTER**.)

9.4 Examining Data

The **Read** menu reappears and the screen work area displays information about the data. This information is recorded, along with the series data values, in the data file for files in the BATS format. As you can see, there are two series in this data set, Sales and Price, and values are recorded monthly from January 1976 to December 1981 inclusive, a total of 72 observations on each series.

▷ Select **Quit** to return to the **Input-Output** menu.

Notice that there are now several further options on this menu in addition to **Read** and **Keyboard** that were all we saw 'on the way down'. These extra options—**Write, Append,** etc.—all require a data set to be present before they represent meaningful activities (think about writing a nonexistent data set to file!) and this explains why these options did not appear earlier. Option suppression is used throughout BATS so that only those facilities which are valid at any stage of a session are available for selection. It is possible that for some analyses one or more of the options will never be available. If you want a sneak preview of the complete option structure, have a browse through Chapter 15.

We will return to these exciting new options shortly but for now let's take a look at the data just loaded.

▷ **Quit** again to the **Data** menu.

There are two new options in the **Data** menu, **Explore** and **Clear**. Highlight each of these in turn (using the left and/or right cursor arrow keys) and read the help bar information.

Let's go ahead with some data exploration.

▷ Select **Explore** and a set of possible data examination utilities is listed.
▷ Select **Table**.

The list display that we previously saw for file selection now appears with the data series names itemised. (If you tried to select **Table** by typing **T** did you follow what happened? Since both **Time-Plot** and **Table** begin with the same letter the choice was ambiguous; a further letter—a for **Table**—was necessary.)

▷ With **PRICE** highlighted press **ENTER**.

A marker appears to the right of **PRICE** but nothing further happens. What has gone wrong? Well...nothing. Another difference between menus and lists is that sometimes it makes sense to choose more than one item from a list. This is *never* the case for menus.

▷ Now highlight **SALES** (press the cursor down arrow key once) and press **ENTER**.

Another marker appears, this time against **SALES**. The markers serve to indicate which list items are currently selected: all of those items marked will be given to whatever procedure initiated the list selection—in this case `Table`.

▷ Press `ENTER` again.

Notice that the marker against `SALES` disappears. `ENTER` works as a toggle in list selection: pressing it sets a marker for the current item if it is not already set, and removes the marker if it is. We want to take a look at both of the series.

▷ Make sure that both `PRICE` and `SALES` are marked for selection.

▷ Press the `ESC` key.

The screen display is replaced by a 'sheet'. Summary information about the data is displayed on the top few lines (the control area) and the selected data series appear in separate columns with column labels at the top and row labels, formed from the data timing information, to the left. The first entry in the table is highlighted, namely `PRICE` for January 1976. You may move around the table using the cursor movement keys and `HOME`, `END`, `Page Up`, `Page Down`. If you experiment with each of these combinations to see how the display changes, you will notice that

- `Home` moves the current position to the top left corner of the existing display page.
- `End` moves to the bottom right.
- `Page Up` displays the previous page (if any).
- `Page Down` displays the next page (if any).

Two other keyboard keys are useful in table displays:

- `F` (upper- or lower- case) allows the table contents—the selected series only, not descriptive or labelling details—to be written to file.
- `?` enables the table display format to be changed.

▷ Press `?`.

Two 'dialogue boxes' labelled `cell width` and `decimal places` appear to the right of the screen in the control area, with a small blinking block cursor in the first. Dialogue boxes are used whenever information such as a name or a date is required. Since BATS has no way of knowing your preferences in such matters it cannot form a menu or list, so you get a 'form' to fill out. Filling out a form entry (a dialogue box) is just a matter of typing the necessary information at the keyboard. Moving between boxes is accomplished using the keyboard arrow keys: down-arrow moves the cursor (or your 'pen' if you like) to the beginning of the next box; up-arrow to the beginning of the previous box. These movements are wrap-around so that from the last box the next box is the first, and vice versa. Try this out for the table format by entering `15` for column width and `5` for decimal places.

9.4 Examining Data 249

▷ Type 15 in the `cell width` box.

Notice that when the cursor reaches the end of a box it automatically moves to the next one in the list. In this case, when you typed the 5 the cursor immediately moved to the `decimal places` box.

▷ Type 5 in the `decimal places` box.

Once again the cursor moved immediately to the 'next' box. Whenever information is being entered into forms like this you can retype box contents and move among boxes as much as you like until you are happy with your entries. When you have finished just press the ESC key to signal completion.

▷ Press ESC.

Notice how the table display is changed. The columns are now wider and additional decimal places are included for the numbers—exactly what we asked for. (Table display format changes are remembered by BATS: the next time you display a table in this session the initial format will be '15/5'.)

Let's now quit from the table and take a look at some graphs, beginning with a basic time series plot.

▷ Press ESC when you have finished with the table, returning to the Explore menu.
▷ Select Time-Plot.
▷ Choose both PRICE and SALES (remember to mark both series as selected by highlighting and pressing ENTER for each in turn) then press ESC.

The standard BATS two panel graphical display appears with a time plot of the PRICE series in the main panel, Figure 9.3, and the legend No Model Set in the summary panel. (Note that the summary panel is not reproduced in the figures in the book.) The graph will remain on the screen until you press Q (or q) to finish the interactive session.

▷ Press Q now.

Another graph appears—this time with the SALES series displayed. (Other data exploration options that take a single series also work in this manner when presented with multiple series: each series is dealt with in order.) If your computer is equipped with a mouse device and a mouse driver is installed (see your DOS manuals for instructions on how to install a mouse driver), then the mouse pointer, typically an arrow, will appear in the upper left corner of the graph. This pointer responds to movements of the mouse device attached to your computer: move the mouse and observe the pointer change position as you do so. The pointer also responds to the cursor movement keys: use with shift pressed for 'large' steps, and without for small steps. Try that too. If your computer does not have a

Figure 9.3 Time plot of price series.

mouse (or if BATS cannot recognise your mouse) a cross-hair cursor will appear in place of the mouse pointer. Move this cursor using the keyboard arrow keys as just described.

Whenever a graph is displayed there is a range of interactive facilities available for adding annotations such as lines or text. Before continuing you might like to experiment: the options are detailed in Chapter 14.

When you have finished:
- ▷ Press Q to quit the graph.
- ▷ Return all the way up the menu tree to the root. (Select Quit at each submenu.)

Notice the new **Model** option in the root menu. This is where we start the next tutorial.

9.5 Ending a Session

Terminating the session is simply a matter of choosing an option from a menu, like all the other operations in BATS.

- ▷ From wherever in the menu hierarchy the program currently is, quit to the root. (If the active menu is a low level menu, you may have to quit through several levels in succession.)
- ▷ From the root select **Exit**.
- ▷ Select **Yes** to confirm the session termination.

Control is returned to the computer's operating system which is restored to the state it was in prior to beginning the session.

9.6 Summary of Operations

1. Starting a session
   - Make sure that your computer's active directory contains **BATS.EXE**.
   - Type **BATS**, then press the **ENTER** key.
   - When BATS appears, press any key: the root menu is activated.
2. Working with menus
   - Select **Data** from the root menu.
   - Select **Quit** to return to the root.
   - Practice with **Configuration** and **Exit** as you like, returning to the root when you have finished.
3. Loading a data file
   - From the root select **Data**.
   - Select **Input-Output**.
   - Select **Read**.
   - Select **Bats-format**.
   - Use the cursor arrow keys to highlight **CANDY.DAT**, then press **ENTER**.
   - Select **Quit** to return to the **Input-Output** menu.
   - Select **Quit** to return to the **Data** menu.
4. Examining data series 1: tables
   - From the **Data** menu select **Explore**.
   - Select **Table**.
   - With **PRICE** highlighted press **ENTER**.
   - Press the down arrow key to highlight **SALES**; press **ENTER**.
   - Press **ESC**.
5. Reformatting tables
   - Press **?**.
   - Type **15** in the column width box; type **5** in the decimal places box.
   - Press **ESC** to reformat the table.
   - Press **ESC** to quit the table and return to the **Explore** menu.
6. Examining data series 2: graphs
   - From the **Explore** menu select **Time-Plot**.
   - Press **ENTER** to mark **PRICE** as selected.
   - Press the down arrow key to highlight **SALES** then **ENTER** to select.
   - Press **ESC** to start the graphs.
   - Press **Q** to quit the **PRICE** time plot; again to quit the **SALES** time plot.
   - Return to the root by selecting **Quit** at each submenu.
7. Finishing the session
   - From the root select **Exit**.
   - Select **Yes**.

9.7 Getting out of Trouble

This is probably the most important section in this part of the book. At least, it may well be during your initiation with the program. By following the instructions here you will be able to recover from any situation that BATS may get you into.

- The ESC key is very useful.

Pressing ESC will complete just about any command in BATS. In menu selection, pressing ESC always returns control one level up in the calling tree. It tells the program that you have no further use for the present menu. In list selection, pressing ESC terminates the selection process. It tells the program that your selection is complete and you would like the items marked to be processed. In dialogue box entry, pressing ESC indicates your response is finished. It tells the program to read your entry (entries if a multiple box form is active) and act on the information received.

- If all else fails, hit the reset button.

This is definitely the refuge of last resort! But it may be your only option if the program really does get into a situation from which you cannot rescue it. Some computers do not have a reset button. On these machines you have to turn the power off then back on to perform a reset. Users of the Windows program should not have to reset the machine. When an unrecoverable situation develops simply kill the window and begin a new BATS session.

9.8 Items Covered in This Tutorial

To round off this introduction to using BATS here is a summary of the points that have been covered.

- Traversing the menu tree; using arrow keys; option selection.
- List item selection including multiple selections.
- Using forms: entering information; moving around forms.
- Loading data files.
- Examining data sets in tables; table reformatting.
- Examining data sets in graphs; interactive additions.
- Conditional option suppression.

Appendix 9.1

Files and Directories

Throughout the program familiarisation in Tutorial 1 we made use of the example data files supplied with your copy of BATS. As you probably already have your own data files—most likely with name extensions different to the ones you have seen used in the tutorial—we shall look here at how you can use your files. We have put this material in an appendix in case you would rather skip directly to the modelling tutorials beginning in the next chapter. Since the examples therein use the supplied data files, you can come back to this material later if you wish.

File Types

There are two aspects to consider: file naming and storage location, and file formats. We will only describe file naming and location here; file formats are discussed in Chapter 13. When BATS is first started it knows about three basic file types:

- Data files containing a data set description and data values: this is the BATS format, and files are assumed to have the file name extension .DAT.
- Data files containing just data values (no information on series names, timing, and so on): this is the so-called 'free' format and files of this type are assumed to have the file name extension .FRE.
- Graphical output files containing PostScript descriptions of graphs produced with BATS; files of this type are assumed to have the file name extension .PS.

The first detail to learn about these file types is that you may choose to name files in any way you wish (subject to DOS limitations) even using file name extensions different from those above. For example, suppose that you have a file of data in the free format (a set of numbers in space separated columns, one series to a column) with file name MYDATA.PRJ. To load MYDATA.PRJ into BATS it is *not* necessary to rename the file MYDATA.FRE (although you may do so if you wish). Select free format read (that is, the following menu selections in turn, starting from the root: Data, Input/Output, Read, Free-Format). The screen displays a list of all those files with file name extension .FRE which are located in the active directory (more about which below). Press Page Down and a dialogue box appears into which you may type MYDATA.PRJ. Press ESC and your file will be loaded. As you can see, the default extension .FRE serves only to identify a base collection of files—it does not restrict the choice of desired file names in any way.

File Masks

Now suppose that you have several free format data files all having the file name extension .PRJ. Wouldn't it be nice if, instead of all those .FRE files appearing, your .PRJ files were listed instead? Then you would not have to bother typing in the full file name each time you wanted to read or write a file (you would not have to remember the file names either!). This file selection criterion can be set by using the Directories entry of the Configuration menu.

▷ From the root select Configuration.
▷ Select Directories.

A set of dialogue boxes is displayed, one group for 'directories' and another for 'file masks'. For the moment you can ignore the directory boxes—we will get to them soon. Notice that the file mask boxes are filled with

*.DAT
*.FRE
*.PS

corresponding to the default file name extensions that we have already discussed. These expressions are known as 'masks' because they serve as a template to which file names are compared. The '*' is a 'wild-card' character which matches 'anything'. So when you request free format data file input, all files which have extension .FRE are included in the list. To change this list to display your .PRJ files, simply replace *.FRE with *.PRJ in the dialogue box (uppercase or lowercase does not matter). To do this, move the cursor to the appropriate box using the up or down arrow keys, then overtype the existing entry. You can move about within a box using the left and right arrow keys, and mistakes may be erased using the backspace key. You might like to practice these motions at this point.

▷ Press the **ESC** key to finish the changes.

If you have changed any of the directory entries and made a mistake, the display will not clear. By 'mistake' we mean entering the name of a directory which does not exist. Either correct the error or, if you do not know what the error is, wipe out the directory entries entirely (use the space bar and/or backspace key to do this). Wiping the entries is equivalent to setting the default startup directory: the directory which was active when you typed **BATS** at the **DOS** prompt to begin the session.

If you entered .PRJ correctly, whenever you request a file in the free format the screen will list your .PRJ files by default. Of course, you may still use **Page Down** to specify some other file name.

The BATS format data file and PostScript graph file masks work in the same way. Moreover, there is no need to use distinct file name extensions to discriminate between any of these file types—but it is well worth doing: consider mistakenly trying to load a PostScript file as a data file!

More information on file masks can be found in your **DOS** manual.

Directories

The 'Directory' boxes which appear along with the file masks are used to set the default directories in which BATS searches for your files. Up to now we have been using the default directory; that is, the directory which was active when you typed **BATS** at the **DOS** prompt to begin the current session. While this is a convenient place to start, it is usually better from a housekeeping point of view to keep similar types of files in their own 'filing cabinet' or directory. Also, groups of files—related to a particular project perhaps—may be kept in separate 'cabinets'. As an illustration, let us suppose that you decide to store your files as follows.

C:\DATA for data files—both BATS and free format.
C:\GRAPH for PostScript graph files.
C:\TMP for work files.

Ignore the last entry for now; we elaborate on it in the next section. With the **Configuration/Directories** option selected you would enter C:\DATA in both the **Bats Format** and **Free Format** directory dialogue boxes; C:\GRAPH in the **PostScript** box; and C:\TMP in the **Working** box. BATS will now search these directories for the files which are listed when, for example, data files are requested. (Only those files matching your file masks will be included in the list of course.)

Remember, if you enter the name of a directory that does not exist, your request will be refused. When you press ESC the screen will not clear. To continue you must enter valid directory names in all three directory boxes. An empty or blank box is considered shorthand for the default directory.

The directory and file mask settings may be changed whenever and as often as you like during a BATS session, although typically this is done only at the beginning. If you set up your directories and file masks so that they differ from the supplied defaults, you will have to reset them each time you begin a new session. You can avoid having to do this if you save the settings using the Configuration/Store option. This action creates a file called BATS.CFG in the default directory, containing your directory/file details. The next time you load BATS from the same directory the configuration file will be read and your settings loaded. Note, however, that if you run BATS from a different directory (that is, one where your BATS.CFG file is not present), then the default file name masks and directories will be set.

The Working Directory

In the discussion above we ignored the Working Directory entry, to which we now briefly turn. During analyses BATS creates and maintains some 'internal' files which are not ordinarily of any use except by the program itself and then only in the session in which they are created. These files have to reside somewhere and the working directory specifies the place. Anywhere will do, but since the files may be discarded after a session a 'scratch' directory is a useful location. BATS itself does not delete these files but existing copies are overwritten every time a new analysis is started, so there is no accumulation of waste. Although you may have different configurations for data files used in different projects, it is usual to have just one common location for working files. The preceding discussion should make clear why this is so. The choice, of course, is yours.

Summary

This appendix has described how to:
- Specify file name masks.
- Specify directory locations for different file types.
- Save your favourite directory and file details.

Chapter 10

Tutorial: Introduction to Modelling

If you have worked through the tutorial introduction to BATS in Chapter 9, then you know how to load data files, display time series in graphs and tables, move around the menu hierarchy, and choose required options. In this tutorial we take the next step: specifying a model for a time series and analysing data using this model.

10.1 Dynamic Modelling

We shall begin with the simplest dynamic linear model, the so-called steady model, or simple dynamic level. Figure 10.1 shows a time plot of the SALES series from the CANDY data set. It is clearly evident that this series will not be adequately modelled by a level component alone, even a dynamic time varying level, because there is strong evidence of systematic seasonal variation. However, without introducing too much detail all at once, using the level model with this series will allow us to demonstrate a number of important modelling and analysis features of the DLM, and the required steps using BATS, as we proceed.

Let us first briefly recap the structure of the steady model, technically the first order polynomial model. The observation and system equations for the steady model have very simple forms,

$$\text{SALES}_t = \text{level}_t + \nu_t, \quad \nu_t \sim N(0, V_t),$$
$$\text{level}_t = \text{level}_{t-1} + \omega_t, \quad \omega_t \sim N(0, W_t).$$

Figure 10.1 Time plot of SALES series from data set CANDY.

In the standard notation for the general DLM the state vector comprises just the level, $\theta_t = \text{level}_t$; the regression vector and system matrix are the constant value 1. This model asserts that the level remains roughly constant over short periods of time but may wander in a random way. The observations Y_t are simply modelled as random fluctuations about this level. Usually the observation variance V_t is much larger than the system variance W_t. Chapter 3 details the formal structure.

For this illustration we will assume that the observation variance V_t is constant but unknown, a situation that is very often found in practice. Also, the system evolution variance W_t will be specified using the discount strategy. (See Part A for discussion of variance learning and component discounting procedures.) Let us now proceed to instruct BATS that we wish to use this model form.

10.2 Specifying the Steady Model

If you have not already done so, load the **CANDY** example data file and return to the **Root** menu. Chapter 9 describes how to do this if you need a reminder. The main menu bar looks like Figure 10.2, displaying options **Data, Model, Configuration,** and **Exit**.

10.2 Specifying the Steady Model

```
BATS 2.10            SELECT              Root menu
Available memory:
[Data] Model Configuration Exit
Data handling and examination
```
title	: CANDY
source	: BATS example dataset
cycle/time	: YEAR/MONTH; period = 12
start : 1976/1; end : 1981/12; length = 72	
2 series:	
SALES	
PRICE	

Figure 10.2 Root menu and display after loading CANDY data file.

▷ Select **Model**.

The screen displays a list of the individual series in the data set, **SALES** and **PRICE**. Whenever a model is defined for the first time after loading a new data set (or having cleared a previous model definition) the response series is the first item which must be specified. Since BATS knows this, it forces the specification as the first step in the model definition phase. Set **SALES** as the response.

▷ Highlight **SALES** using the cursor movement keys.
▷ Press **ENTER**.

Only a single series may be selected for the response, so the list display is replaced by the model menu immediately upon making a selection by pressing **ENTER**.

▷ Select **Quit** and return to the root.
▷ Now select **Model** again.

Notice that this time the model menu appears directly: since we have already set the response series it is not necessary to do so again. (Do not worry about the possibility of having defined the wrong response series: we will shortly see how it may be reset.) There are two options, **Components** and **Interrupts**, on the model menu. We are concerned here with completing the specification of the model form.

▷ Choose **Components**.

(We will get to **Interrupts** in a later tutorial, for now simply ignore them.)

The model components menu contains an entry for each of the possible component forms. The names should be self-explanatory but if you highlight each entry in turn, you can see a brief description on the menu help bar. At the moment we are interested only in the **Trend** entry. (You may also be interested in the **Response** entry at this point if the initial setting made above, on first selecting **Model**, was incorrect. If you need to reset the defined response series, select **Response** and proceed to choose **SALES**.)

▷ Select **Trend**.

The **Trend** component menu allows either a constant or a linear trend component to be included in the model. One or the other *must* be selected: as far as BATS is concerned a model is not defined until a trend component, in addition to the response, is specified. For the steady model we need to select a constant trend.

▷ Select **Constant**.

10.3 Prediction

The model form is now complete. The response series, y_t, is **SALES** and a dynamic level model form is set. We are now ready to perform a time series analysis: we can calculate predictions from the model, or go directly to 'fitting' the data. To begin, let us do some prediction from the steady model for **SALES**.

▷ Return up through the menu structure to the root.

You will see that there are two options which have not appeared before, either here or in the previous tutorial, **Fit** and **Predict**. For prediction (you guessed!)

▷ select **Predict**

and two dialogue boxes for the level appear. These are for setting the initial conditions for the prediction: so far we have a model *form* but no idea of its *quantification*. These two boxes are for setting the initial value of the level—its mean and standard deviation.

▷ Enter 100 for the mean and 1 for the standard deviation.
▷ Press **ESC**.

Another pair of dialogue boxes appears, this time to quantify the observation error term ν_t.

▷ Enter 1 for the estimate and 1 for the degree of freedom.
▷ Press **ESC** to set the entered values.

Next we have to specify how far ahead we wish to forecast—remember that the current standpoint is the beginning of the data series, so we will

10.3 Prediction

be forecasting from 1976/1 onwards. The screen displays a summary of the data set and allows the *forecast horizon* to be set. You can see from this information that **CANDY** is monthly data so that 1976/1 is January 1976 and so on. We will calculate forecasts for the whole of the first year so the required forecast horizon is December 1976 or 1976/12. The month and year are actually in separate dialogue boxes (the dividing slash is not in either box and is therefore not affected by what you type) so we can move directly to the month box and change the 1 to a 12.

▷ Press the cursor down arrow key to move the cursor to the month box.
▷ Enter 12.
▷ Press **ESC**.

This will give forecasts for the first year of the data, based on the initial conditions we have just defined for the steady model. BATS is now ready to get on with the prediction calculations but before doing so we have an opportunity to review—and redo if necessary—the initial value settings just made. Try this if you wish (you can do so as often as you like before deciding to continue), but note that the following discussion assumes the values given above. To get on with the calculations,

▷ select **Continue**.

Once the prediction calculations are complete, a graph of the results is displayed. This forecast graph contains several pieces of information:

1. The observation series **SALES** for the forecast period.
2. The forecast mean.
3. 90% uncertainty limits about the forecast mean.

Take a look at these items in turn. The values of the **SALES** series are quite different from the predicted mean (much smaller as you can see). This is a direct result of the prior mean value for the level which we set at 100. We deliberately chose such an erroneous value to demonstrate that predictions are based entirely upon the model form and its initial quantification: predictions are only as good as the information that is used to make the calculations. You will have often heard the phrase 'garbage in, garbage out'—translated into language more appropriate for the present context this just means that if you start from the wrong place you are likely to end up in the wrong place even if your 'path' (model) is appropriate. Only a fortuitous combination of other errors could change this outcome!

The forecast means for 1976 are constant at the value 100—the prior mean for the level. This is precisely the form of the forecast function for the steady model—a constant. Notice, though, the pattern of divergence in the uncertainty limits. As we project further into the future the predictions become progressively more circumspect. Why is this? The reason is the dynamic nature of the model: a nonzero evolution variance increases

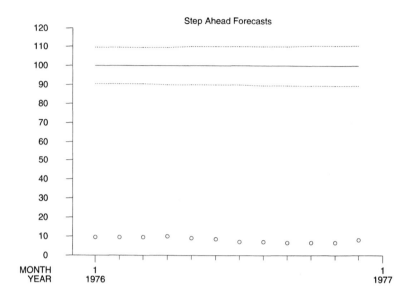

Figure 10.3 Predictions for 1976, steady model for SALES series.

uncertainty about the level each time we move forward. The predicted level estimate itself does not change without information external to the model but our uncertainty increases. We glossed over this issue of temporally induced, or evolution, uncertainty in the model quantification preceding the calculations—in fact we ignored it entirely—so BATS went ahead with its default action. Evolution covariances are determined using the block discounting strategy and to calculate them we have to set values for component discount factors. The steady model has only a trend component so a single (trend) discount is required and in the current illustration the default value of 0.9 was used because we did not specifically set one. In effect, about 10% of the information on the level is 'lost' each time we move forward, hence the diverging limits in the graph, Figure 10.3. If we now repeat the prediction but explicitly set the discount to 1, so defining a static model, then the prediction uncertainty will be constant no matter how far into the future we forecast. We will see this in the next section.

When you have finished examining the graph

▷ press Q to quit.

The **forecast** menu appears. Here you may examine the forecast values in a table (or redisplay the graph), and even save the table to file if you wish. Recall from Tutorial 1 that writing a table to file is initiated by pressing **F** while the table is displayed.

10.3 Prediction

At this point let us summarise what we have done in this tutorial so far.
- Specify the model response series, SALES.
- Specify the model form, steady trend.
- Calculate predictions for SALES using the defined model and a particular set of initial conditions.
- Examine predictions by graph and by table.

10.3.1 Changing the Model Dynamic

Very often when analysing what might be going to happen in the future we will have a variety of alternative scenarios to consider and will want to compare predictions for each of these. The prediction facility just illustrated may be used as often as you like, whether before model estimation as we have been doing up to now, or part way through analysing a series, or after analysing an entire series, and making whatever assumptions (that is, setting initial conditions) necessary for alternative views. As an example, let us repeat the prediction we did above, but this time we will explicitly set the trend component discount factor.

▷ First, return to the root: select Quit at each submenu.
▷ Select Model.
▷ Select Discount.

The discount menu options are to set a single discount factor for individual model components that will apply for all times, or to set each component's discount factors individually for each time. We want to change the trend discount factor from its default value of 0.9 to the static model value of 1.0 and have this apply to every time point.

▷ Select Constant.
▷ Enter 1.0 for the trend.

Ignore the variance discount for now: it plays no role in prediction. (Incidentally, while we are in the model menu, have you noticed the Remove entry? Once a model is defined, which means the response series and a trend component are set, you have the option to wipe it out in one go. But don't do that just now!)

▷ Press ESC.
▷ Return to the root.

If you now repeat the prediction procedure we went through above, a similar forecast graph will appear, with one major difference in content. This time the uncertainty limits do not diverge as we forecast further ahead because in the static model (from setting a unit discount) there is no inherent temporal variability in the system evolution. Today's level is known precisely as it was yesterday, and the day before that, and so on.

You may like to try a few more variations on this prediction theme. Try changing the prior values for the level and observational variance for example, or choosing a different forecast horizon. (The 'Summary of Operations' list, item three, summarises the necessary commands. Just substitute your prior and horizon values for the ones we used here.)

10.3.2 Modelling Known Variance

BATS does not include an option to explicitly fix the observation variance, V_t, at a known value. However, the same effect may be simulated by setting the degrees of freedom parameter to a very large value—say 10000—and the variance component discount factor to unity. To all intents and purposes the variance is then fixed at your specified point estimate. Why not repeat the predictions we did above, this time with V_t 'fixed' at the estimate 1, and compare the forecast uncertainties with the previous values. (Once again, refer to the 'Summary of Operations' item three for the required steps.)

10.4 Forecasting and Model Estimation

Having seen 'pure' prediction based on a given quantification of the steady model, let us now turn to applying the model to the SALES data. We start once again from the root. To begin, we shall restore the default trend component discount factor.

- ▷ Select Model.
- ▷ Select Discount.
- ▷ Select Constant.
- ▷ Enter 0.9 for the trend.
- ▷ Press ESC.
- ▷ Quit to the root.

Now to get on with the analysis.

- ▷ From the root select Fit.

Just as we did for prediction we must determine the starting conditions for the analysis. In this case though we have an option to let BATS determine some starting values for us—the so-called reference analysis. This is perhaps most easily thought of as saying, 'We don't have any real idea about SALES so go ahead and observe the series for a while, then, when you have a picture of the series form, assign some 'reasonable' values to the model parameters to reflect this picture in the model. But also reflect great uncertainty'.

10.4 Forecasting and Model Estimation

We begin by using reference initialisation.

▷ Select **Reference**.

Observe that the time frame display appears, complete with a summary of the model definition in the right hand panel. The **SALES** series is also shown, each value being drawn along with its forecast as the latter is calculated. This dynamic display update is extremely useful when we want to perform interventions, that is, suspend the analysis at some stage and alter the model quantification before continuing—a change in level for example—or examine forecasts for some period following the break, and so on. We will see how to perform these intervention analyses in the next tutorial.

Notice that the line drawn through the series joins one step ahead point forecasts. In particular, it is important to realise that *these values are not fitted values*. The point forecast for today's sales is the expected value of the distribution of sales based on the assumed model form using parameter values deemed most appropriate given the behaviour of sales up to yesterday, $E[Y_t|D_{t-1}]$.

When the analysis is complete—the value of **SALES** for December 1981 has been drawn—you may quit from the graph by pressing **Q**. There are no interactive facilities available for this graph so no mouse pointer or crosshair cursor appears.

▷ Press **Q** to quit the forecast graph.

When you quit from the forecast display the root menu is active once more, with yet another new option available, **Retrospective**. The analysis just performed is the 'on-line' analysis or 'forward filtering', which is to say that the data have been analysed and forecasts produced as they would be in a routine 'live' situation—with no benefit of hindsight. Model fitting, or estimation with hindsight, is known as 'back filtering' or retrospective analysis. The **Retrospective** option calculates these values as we will see shortly.

Before we proceed with further analysis, take a look at some of the results we already have.

▷ Select **Data**.
▷ Select **Explore**.
▷ Select **Time-Plot**.

The list of model estimates and data series is now displayed. As you did in Tutorial 1, any combination of these series may be selected—multiple choices are possible if you recall—and each selected series will be graphed in turn. In the list, estimate series prefixed with a * will include the response series in the graph; the O prefix denotes 'on-line' results, the significance of which will become apparent later. Select some of the estimate series

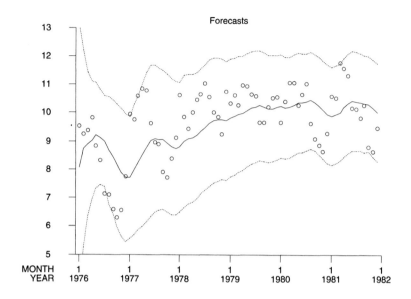

Figure 10.4 One step forecasts for SALES series.

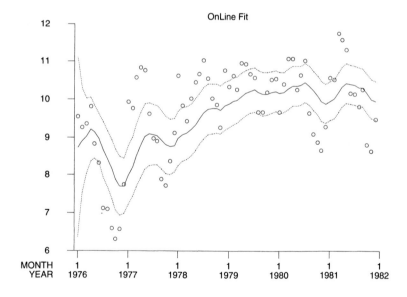

Figure 10.5 On-line fitted values for SALES series.

10.4 Forecasting and Model Estimation

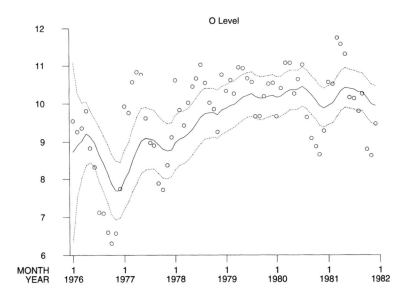

Figure 10.6 On-line estimated trend for SALES series.

(and data too if you like) and examine the graphs. Figures 10.4, 10.5, and 10.6 show the one step forecasts, on-line fitted values, and level series respectively.

We have already discussed the forecast series. The on-line fitted value for today's sales is the model's best estimate of this value given the data up to and including today's observed value, $E[\text{level}_t | D_t]$. This best estimate is typically not equal to the actual value, even though this has been observed and included in the estimate, because the model is constrained to provide best estimates for the values of sales yesterday and days before in addition to today. The on-line estimated level is identical to the fitted value because the steady model is composed only of the single parameter trend component. In more complex models these two series will be quite different as we will see later in this tutorial.

10.4.1 Retrospective Analysis

When you have finished examining the on-line results we can go ahead with the smoothing analysis.

▷ Return up the menu hierarchy to the root.
▷ Select **Retrospective**.

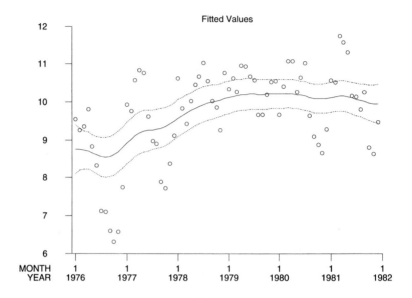

Figure 10.7 Retrospective fitted SALES.

In contrast to the way the forecasts were produced above, the back filtered estimates are computed in reverse order, from the end of the series to the beginning. The values being calculated here and shown on the graph (Figure 10.7), are the model's best estimated values for sales for each time *given all of the observed values*. The distinction between these fitted values and the on-line fitted values is extremely important for assessing 'what really happened' in the data. Where the on-line fitted value for, say, March 1979 is calculated from the standpoint of the end of that month, the retrospective fitted value is calculated with the hindsight of what also was observed in the SALES series from April 1979 to December 1981.

Take a minute to think about this distinction: you need to keep in mind both the form of data and the structure of the assumed model. In January 1976 we are quite unsure about the level of sales. Our best guess turns out to be in the right ball park but our uncertainty is really very great and we must rely quite heavily on the information coming from the sales data as it becomes available to reduce this uncertainty and improve our estimate. Early on then the estimated level—this is the only component in the steady model—adjusts to follow the pattern in the data. Moving through the second year, model adjustments to the wave-like pattern in sales are rather less than in the first year. The basic steady model form prefers less movement to more—none at all in the static case—and we have

10.4 Forecasting and Model Estimation

learned from 1976 that over a full year up and down movements in sales tend to average out.

But beware of falling into the trap of thinking that this development will eventually mean the estimates will become constant. They will not. The reason they will not is that a dynamic model pays less attention to observed values as they age (if the discount factor is less than one). In this case, for reference initialisation, the smoothed estimate is a weighted average of observations either side of a given time where the weights decay exponentially, $\sum_{i=1-t}^{T} \delta^{|t-i|} y_{t-i} / \sum_{i=1-t}^{T} \delta^{|t-i|}$. A useful exercise that you might like to try later is to generate some data from a pure sinusoidal function and fit the steady model with discounts 0.9 and 1.0. Explore what happens to the on-line estimates as more and more data are analysed. Also try a discount of 0.5 and watch the estimates lag behind the observations.

The retrospective analysis has the full advantage of hindsight, the big picture if you will. It is able to consider the data for all six years and make the best overall compromise between the variability in sales and the underlying assumed model form. Had we set the trend discount to unity for a static model the retrospective estimate would have been restricted to finding a single constant value for the level. This single value would have been the best compromise between the average sales value and our initial guess. With a discount factor less than 1 the fitted estimates can vary over time. Think of an analogy of placing a stiff but nonrigid strip of plastic through the data as plotted so that it gets as close as possible overall to the plotted points. Smaller discount factors correspond to more pliable plastic which is more easily shaped. In the limit, with the discount approaching zero, the plastic becomes infinitely elastic and can be bent to fit through every point exactly. At the opposite extreme a discount of 1 means that the plastic cannot be bent at all from its straight line shape. In the analysis here you can see that there is indeed quite strong evidence of movement in the underlying level of sales, although until the other clearly evident structure in the series is modelled it would be premature to make much of this.

As with the forecast graph there is no interactive facility with the retrospective graph. Quitting from the graph returns control to the root menu.

▷ Press **Q**.

If you now select **Data**, **Explore**, **Time-Plot**, the list of data and estimate series will be displayed. This time both the on-line and retrospective estimates are included in the list. The 0 prefix indicates on-line estimates, and S indicates 'smoothed' or retrospective estimates.

An informative view of the different model estimates is obtained by using the **MPlot** option of the **Explore** menu. This option displays series in sets of up to three on a page which is useful for making comparisons. If you use

MPlot to graph the forecasts, on-line fit, and fitted values, the distinctions described above can be seen very clearly in the three-way display.

- ▷ Select **Mplot**.
- ▷ Using the arrow keys highlight **Forecasts**, **OnLine Fit**, and **Fitted Values** and press **ENTER** for each in turn.
- ▷ Press **ESC**.

When you have finished examining the graph quit to the root.

- ▷ Quit the graph.
- ▷ Quit to the root.

10.5 Summary of Operations

1. Loading the CANDY data set
   ▷ From the root, select Data, Input-Output, Read, Bats Format, CANDY.
   ▷ Quit successively up the menu hierarchy to the root.
2. Specifying the steady model
   ▷ From the root, select Model, SALES, Components, Trend, Constant.
   ▷ Quit to the root.
3. Predictions
   ▷ From the root, select Predict.
   ▷ Enter starting values for the mean and standard deviation of the level (100, 1); press ESC.
   ▷ Enter starting values for the observation variance and degrees of freedom (1, 1); press ESC.
   ▷ Enter the forecast end time (horizon) 1976/12; press ESC.
   ▷ Select Continue.
   ▷ Q to quit the graph.
   ▷ Quit to the root.
4. Changing discount factors
   ▷ From the root, select Model, Discount.
   ▷ Select Constant to set the same component discount factors for each time.
   ▷ Enter 1.0 for the trend; press ESC.
   ▷ Quit to the root.
5. Data analysis 1: one-step forecasting
   ▷ From the root, select Fit, Reference.
   ▷ Quit the graph.
   ▷ Select Data, Explore to examine on-line estimates.
   ▷ Quit to the root.
6. Data analysis 2: model fitting (retrospective analysis)
   ▷ From the root, select Retrospective.
   ▷ Select Data, Explore to examine on-line and smoothed estimates.
   ▷ Select series by highlighting and pressing ENTER.
   ▷ Press ESC when all required series are marked for selection.
   ▷ Quit the graph.

10.6 Extending the Model: Explanatory Variables

With sales data it is very often found that a relationship exists between sales volumes and prices. This is intuitively 'obvious' and there are many economic theories to explain such a relationship, and why different market structures result in different relationships. The example data set contains unit prices in addition to the sales series on which we have concentrated attention so far. Begin by taking a look at the sales and price series together.

▷ Quit successively to the root.
▷ Select **Data**.
▷ Select **Explore**.
▷ Select **Mplot**.
▷ Mark both **SALES** and **PRICE** for selection (highlight each in turn and press **ENTER**).
▷ Press **ESC**.

Both series are graphed (Figure 10.8). There appears to be some evidence of coordinated movement between sales and price: price increases having a tendency to be associated with sales decreases. However, the time plot comparison is a little difficult to interpret because of the seasonal movement in sales which we have already remarked on. A direct plot of sales against price might be more helpful.

▷ Press **Q** to quit the current graph.
▷ Select **X-YPlot**.
▷ Choose **SALES** and **PRICE** as above.
▷ Press **ESC**.

This correlation plot (Figure 10.9) shows up quite a strong linear relationship between the variables—with correlation -0.63—and negative as expected.

It would certainly seem from such cursory examination that we can build a better model for sales by including a relationship with price.

10.6.1 Specifying a Regression Component

The existing model definition comprises just a constant trend term. (If you have erased the model used in the previous analysis, or are starting this example from scratch, you will need to specify the response series, **SALES**, and a constant trend component. See the preceding 'Summary of Operations', part 2, for a reminder of the necessary steps.) We now want to add a regression on the price variable:

$$\text{SALES}_t = \text{level}_t + \beta_t \text{PRICE}_t + \nu_t.$$

10.6 Extending the Model: Explanatory Variables

Figure 10.8 Multiple time plot of SALES and PRICE.

(This model is discussed in Part A, where the standard DLM representation in terms of the usual 'abstract' quantities, y_t, θ_t, and so on, is also given.) Notice that the coefficient of price, like the trend, is time indexed—a dynamic parameter. The model system dynamics are extended to include specification for this regression coefficient so that we now have

$$\text{Level}_t = \text{Level}_{t-1} + \omega_{1t},$$
$$\beta_t = \beta_{t-1} + \omega_{2t}.$$

Just as the level is allowed to adjust with time, so is the influence of price on sales.

We add a regression on PRICE to the model in a way similar to how the response series was set.

▷ Quit to the root.
▷ Select **Model**.
▷ Select **Components**.
▷ Select **Regression**.

In the list of displayed series from which to choose regressors notice that **SALES** does not appear. BATS knows that it is not a sensible idea to regress the response series on itself!

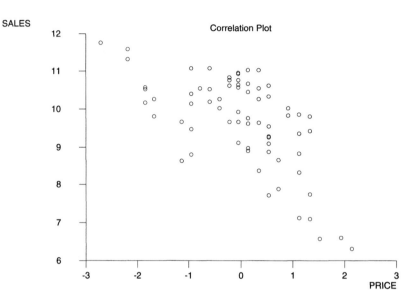

Figure 10.9 X-Y Plot of SALES and PRICE.

▷ With PRICE highlighted press ENTER, then ESC.

The model now includes a dynamic regression on price as well as the dynamic level. If you graph a series (Data, Explore, Time-Plot) you will see the model description in the summary panel includes the regression on price.

Let us go ahead and fit this new model to the sales data using the reference initialisation.

▷ Quit successively to the root.
▷ Select Fit, then Reference.

The one step ahead forecasts show quite a lot of improvement for much of the data over the forecasts from the level only model (compare the current graph with Figure 10.4.). As we expected, the price variable does indeed appear to have information useful in predicting sales volume. To get a more complete picture of this analysis perform the retrospective fit, then go to the estimate exploration menu.

▷ Quit the forecast graph.
▷ Select Retrospective.
▷ When the calculations are complete, quit the retrospective graph.
▷ Select Data then Explore.

The list of estimate series is more extensive than we saw for the analysis

10.6 Extending the Model: Explanatory Variables

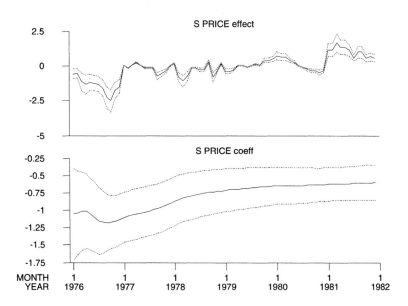

Figure 10.10 Smoothed PRICE coefficient and effect.

of the steady model. There are now entries for the regression component: one for the coefficient estimate, $E[\beta_t|Data]$, and one for the effect of price on sales, $E[\beta_t|Data] \times PRICE_t$. There are series for both the on-line and smoothed estimates.

Examine the smoothed price coefficient and effect series with the MPlot option (Figure 10.10).

▷ Select MPlot.
▷ Highlight S PRICE Coeff, then press ENTER.
▷ Highlight S PRICE Effect, then press ENTER.
▷ Press ESC.

The coefficient estimate and 90% uncertainty limits are displayed in the lower frame of the graph. As you can see, the estimated mean value is not constant over time, showing an overall increase from about -1 at the start of the series in January 1976, to about -0.6 at the end of 1981. This movement is allowed by the dynamic model with a regression discount factor less than unity. The default value is 0.98 and since we did not explicitly set discounts in this analysis, that is the value used here. Note that this value is much larger than the default trend discount which is 0.9. (The general component discounting strategy is discussed in Chapter 2 and Chapter 3.)

These relative discount values reflect the general observation that regression relationships tend to be more stable than pure time series components such as trend and cyclical variation. Discount factors will often be different from the default values that we have been using for convenience—remember that smaller discount factors mean greater uncertainty so we always strive to maximise discounts while allowing for whatever movement is necessary to reflect actual market dynamics and modelling approximations. Although discount values may be changed, the ordering of 'regression discount greater than trend discount' will usually be appropriate. Remember also that with a regression component defined, the level is effectively playing the role of regression intercept. Bearing this interpretation in mind there is a strong argument for the level discount to be quite close to the regression discount. For an example where trend and regression components exhibit quite different amounts of dynamic movement, and wherein the conventional wisdom espoused here of regressions more stable than trends, see the market share application in Chapter 5.

In practice one would compare a range of discount factors for each of the model components and explore comparative forecasting performance. Every series is different, some have greater stability in one component, while others have greater stability in other components. And of course there is the crucial issue of periods of greater or lesser stability, or structural change. The search for a single set of optimal discount factors applicable for all time is often a search for the pot of gold at the end of the rainbow. The goal is impossible to achieve. The next tutorial delves into the issues arising from exceptions and changes in temporal stability.

The upper frame of the graph in Figure 10.10 shows the contribution of price to the estimate of sales, namely $PRICE_t \times E[\beta_t|Data]$.

Now take a look at the on-line price coefficient estimate and compare it with the smoothed estimate.

▷ Quit the current graph.
▷ Select **MPLOT**, choose **O PRICE coeff** and **S PRICE coeff**, then press **ESC**.

Figure 10.11 reproduces the graphs. Notice how much more stable the retrospective estimate is, and the reduced uncertainty—be careful in interpreting the vertical scales on the two frames: they are *not* the same. The uncertainties may be more easily compared in tabular form (**Table** option on the **Data/Explore** menu). At this point you may want to review the discussion of on-line and retrospectively smoothed estimates from the steady model analysis.

Have a look at the on-line and smoothed estimates of the level, reproduced in Figure 10.12 (**MPlot** again, this time with **O Level** and **S Level**). The

10.6 Extending the Model: Explanatory Variables

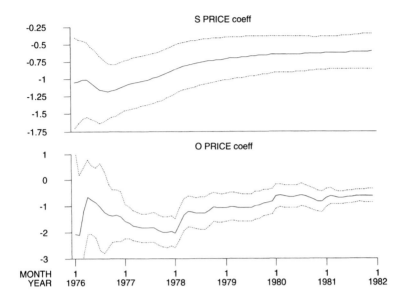

Figure 10.11 On-line and Smoothed PRICE coefficient.

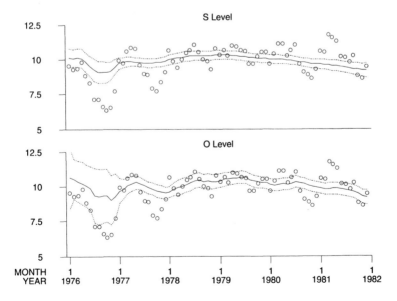

Figure 10.12 On-line and Smoothed level.

estimates are much more stable and less uncertain than the estimates from the steady model (Figures 10.6 and 10.7). In that simpler model the level was trying to be 'all things to all data'. The movement in sales that the current analysis shows directly related to movements in price had to be reflected in erroneous movement in the level: there was nowhere else for it to go. (By the way, don't be confused by the labelling of Figure 10.7 as 'fitted values'. For a level only model, the fitted values are identical to the estimated level.)

When you have finished return to the root menu.

▷ Quit the graph.
▷ Quit successively to the root.

Correlation and Regression Coefficients

To round off the present discussion consider the final posterior estimate of the regression coefficient, -0.6. If you perform the analysis for a static model (all component discount factors equal to unity) the estimate is -0.76. Neither of these values equals the correlation coefficient that we observed in the X-Y plot at the start of this section, -0.63. In fact the static analysis is substantially different from this latter value. To understand the differences here you have to remember how correlations are defined. In particular the scaling of variables is irrelevant because they are standardised in the computation of the correlation coefficient. But this is *not* true for the regression model we have analysed. The PRICE variable is in fact standardised but SALES is not. And therein is the cause of the estimate and correlation differences. If a standardised sales series is used in the level plus regression model of this section we would indeed observe that for a static model the regression coefficient is precisely the linear correlation value.

10.7 Data Transformations: New Series from Old

BATS provides a built-in facility for creating new time series as transformations of existing series. Simple transformations including logarithms (natural and decimal), exponentials, power transforms, centreing, and standardisation may be applied to series using the Create option of the Input-Output menu. A full list of the available transformations is given in Chapter 15. Let's go ahead and check out the claim made at the end of the previous paragraph. To do this we need to construct a standardised version of the sales series.

BATS should currently be displaying the root menu. If it is not, quit successively until it is.

▷ Select Data, Input/Output, then Create.

10.7 Data Transformations: New Series from Old

Two dialogue boxes appear. Do you recall the nature of these boxes from the prior specifications we did when forecasting from the steady model? Well, the present boxes operate in a similar manner except that we will be entering text rather than numbers. The first box, labelled 'target series:', is for the name of the new series that we are going to create. You can choose whatever names you like for new series (subject to the name length limitation implicit in the dialogue box length). For standardised sales let's be very creative and use **stdsales**. Type case is significant so if you want to maintain compatibility with the existing series and capitalise the name be sure to enter the name as **STDSALES**. It is your choice. By the way, if you enter the name of a series already in the data set, BATS will ask you to confirm that you wish to overwrite the existing values. Be careful with this feature. Once a series is overwritten there is no way to retrieve it.

▷ Enter **stdsales** in the target series box.

Pressing the cursor down arrow key will move attention to the second box. If you press down arrow repeatedly you will see the small blinking block cursor move from one box to the next, cycling back to the first when the last is reached. You might also hear a 'beep' as the cursor moves. The second box, labelled 'formula:', is for entering the required transformation (see Chapter 15 for a list of available transformations).

▷ Use the down arrow key to move the cursor to the formula box.
▷ Enter **std SALES** (you may use any mix of uppercase or lowercase here).

BATS maintains the case of the name that you enter as the target, but recognises any mix of lower and upper case when searching the list of known names. So you could have entered 'std sales' or 'std Sales' and so on in the formula box to achieve the same effect here.

When you have entered the target series and formula information, you need to tell BATS that you have finished and would like the specified series created.

▷ Press **ESC**.

The new series will be created and entered into the current data set. Note, however, that the modified data set will not be updated on file unless you explicitly write it (the **Write** option on the **Input/Output** menu). The data set summary now displayed should include the new series in the name list. Let's see what this series looks like.

▷ Quit from the **Input-Output** menu.
▷ Select **Explore**, then **MPlot**.
▷ Choose both the original **SALES** series and the new **stdsales** series (highlight and press **ENTER** each in turn).
▷ Press **ESC**.

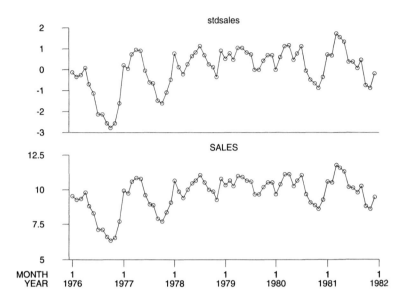

Figure 10.13 Timeplot of SALES and stdsales.

As you can see in Figure 10.13, both series have exactly the same form, differing only in their measurement scale.

You can now replace the model response series SALES with the standardised series stdsales and repeat the analysis above. We leave this as an exercise for you. (The next 'Summary of Operations' items 2–4 lists the steps we followed if you do not want to re-read the previous section.)

10.8 Prediction with Regressors

Making predictions from models which include regressions on external variables is a more complicated matter than we saw with predicting using the steady model. The problem is that we have to determine values for the regressor variables over the prediction period. In real situations this is often one of the most difficult aspects of forecasting because the future values of external variables are themselves uncertain. One way in which this uncertainty is handled is to calculate sets of predictions for a variety of possible situations, corresponding to different possible outcomes of the external variables. Chapter 5, Section 5.6, discusses this matter further.

10.8 Prediction with Regressors

With the difficulty in mind, given future values of external variables in a model it is a straightforward matter to produce forecasts in BATS.

The basic procedure is similar to what we went through with the steady model, except that when the forecast horizon lies in the future (later than the last recorded time point of the data set) we have to set future values for the model regressors. If you have followed the analysis of the regression model above, then a current analysis exists and the prediction option (from the root) will take the standpoint of forecasting from the last time point of the data.

▷ Select **Predict** from the root.

▷ Enter **1982/12** for the forecast horizon.

The first forecast is for the first month following the data set as we just said. This is different from what we did previously where we were forecasting before doing any analysis and the starting point was the beginning of the data.

▷ Press **ESC**.

The table display appears with the data series displayed by column. This may seem a strange thing to happen given that you have requested prediction, but there is good reason!

▷ Press **Page Down** repeatedly to go to the end of the data set.

You can see that the values of the series for 1982, the year for which we have requested predictions, are all zeroes. BATS knows that for models which contain regressors, prediction beyond the end of the data set will require that values for the regressors be entered; hence the table. For this example we can leave the zeroes so that price will have no effect on the forecasts.

▷ Press **ESC**.

▷ Select **Continue**.

See how the predictions follow the form of the steady model: a straight line (Figure 10.14). This is precisely what we expect for a constant price. Moreover, with price set to zero the forecast means will be equal to the final posterior level mean (9.211, as you can see from the forecast table if you quit the graph). To see the effect of setting different prices for 1982 go through the prediction steps again but this time edit the values for price in 1982. We leave you to experiment with this.

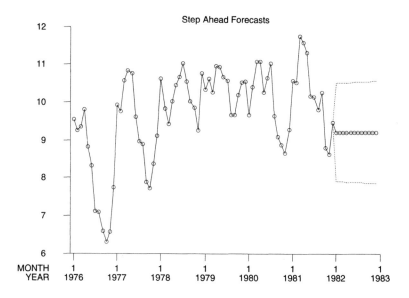

Figure 10.14 Predictions.

10.9 Multiple Regressions

Regression components are not restricted to single variable regressions. A model may contain as many regression relationships as you wish (within the physical restrictions of the program and computing environment!). When selecting regressors—as we did for PRICE in a previous section—you highlight each required regressor variable in turn and press ENTER. This marks the variables for selection. A subsequent pressing of the ESC key causes BATS to set the (multiple regression) component definition. The selection procedure is identical to the estimate selection procedure that we have already seen on several occasions in this tutorial.

Each regressor variable is treated in the same way as the single variable case. Prior moments may be specified, and individual regression parameters and effects may be examined following an analysis. And note that the regression component discount factor is common to all regressions. Regressions are grouped into a single block component for discount evolution. (Chapter 5, Section 5.8, discusses multiple regression blocks and separate discounting of individual regressors.) There is also something extra. An overall regression component estimate—the sum of the individual regressions—labelled 'regression block' is calculated and included in the estimate list.

10.10 Summary of Operations

1. Correlation Plot
   - From the root, select **Data, Explore, X-Y Plot**.
   - Highlight **SALES**, then press **ESC**.
   - Highlight **PRICE**, then press **ESC**.
   - Press **Q** to quit the graph.
   - Quit successively to the root.
2. Defining a regression (with response and trend already set)
   - From the root, select **Model, Components, Regression**.
   - Highlight **PRICE**, press **ENTER**, then **ESC**.
   - Quit to the root.
3. Fitting the model
   - From the root, select **Fit**, then **Reference**.
   - Quit the forecast graph.
   - Select **Retrospective**.
   - Quit the retrospective graph.
4. Examining analysis results
   - From the root, select **Data, Explore, MPlot**.
   - Highlight **S PRICE Coeff**, then press **ENTER**.
   - Highlight **S PRICE Effect**, then press **ENTER**.
   - Press **ESC**.
   - Repeat and/or use **Time-Plot** for displaying any other series in the list.
   - Quit to the root.
5. Data transformations—standardised **SALES**
   - From the root, select **Data, Input-Output, Create**.
   - Enter **stdsales** for the target series.
   - Enter **standardise** for the formula.
   - Press **ESC**.
   - Quit to the root.
6. Prediction following analysis (assumes on-line analysis completed)
   - From the root, select **Predict**.
   - Enter forecast end time, 1982/12 for a full year following data; press **ESC**.
   - Press **ESC** to quit table (leaving future regressors at zero).
   - Select **Continue** to use pure model based priors, or **View-prior** to examine them first, **Edit-prior** to reset them.
   - Quit from the graph
   - Quit to the root.

10.11 Extending the Model: Seasonal Patterns

We have remarked on a couple of occasions in this tutorial that the sales data appears to have a fairly strong pattern of seasonal variation over each year. But so far our modelling has ignored this observation. It is time to redress the omission.

Seasonal patterns of arbitrary shape may be modelled within the dynamic linear model. A seasonal component is defined by the period over which it cyclically repeats and the particular form of the variation over that period. Such a component is similar to the trend component in that it is a pure time series component—no external variables are involved. It has a mathematical structure that provides a range of behaviour restricted by repetition after a finite number of steps. The mathematical details are set out in Chapter 3.

10.11.1 Specifying a Seasonal Component

The response series, the trend, and a regression on the PRICE variable are already defined, so we just add a seasonal component to the existing model.

▷ Quit to the root.
▷ Select Model.
▷ Select Components.
▷ Select Seasonal.
▷ Select Free-form.

A free form seasonal pattern means a pattern of arbitrary shape over the cycle period. The alternative Restricted-harmonic specification provides for parsimonious seasonal modelling (when it is appropriate): we investigate restricted seasonal patterns in the next section so you can ignore this for now. The free form option assumes that the cycle period is the same as the period of the data, which is 12 in the current example. The sales data is monthly and has an annual seasonal cycle in common with many business series so the period length of 12 is fine. Should you wish to set a seasonal period different from the data set period you may do so by travelling the Restricted-harmonic path; again, we refer you to the next section to see these specialisations.

Go ahead and fit the model; as with our previous analyses we use the default reference initialisation.

▷ Quit to the root.
▷ Select Fit, then Reference.

Don't be surprised that the analysis is rather slower than we have become used to up to now. By adding an unrestricted seasonal component to

10.11 Extending the Model: Seasonal Patterns

the level plus regression model we have increased the model parameter dimension from 4 to 15. (The extra 11 parameters correspond to 12 seasonal factors with the constraint that they sum to 1 over a full year.) Basically this means that there is a lot more computational work to do.

When the on-line analysis is complete, quit the graph and go immediately to the retrospective analysis. We will look at both sets of results together.

▷ Quit the forecast graph.
▷ Select **Retrospective**.

The back filter calculations also take rather longer as you can see. Before you quit the retrospective graph take a look at how well the model now fits the data. Compare Figure 10.7 for the simpler nonseasonal model that we have previously analysed. An improved fit was bound to occur as a direct result of the greater number of parameters in the model, but are the parameters really useful? Questions we need to ask at this point include:

Have we got the right parameters (model structure)?
Is there redundancy in the model?

These questions (and others) become more important as the model complexity increases. Getting a good model fit is 'easy' with sufficient parameters almost regardless of the model structure. But answering questions like the foregoing is crucial in determining whether we really have a good explanation of the data, and therefore whether the model based predictions will be useful.

Let's take a look at the results of the analysis.

▷ Quit from the retrospective graph.
▷ Select **Data**, then **Explore**.

Do you notice the two new options, **Seas** and **Factor**? This is another example of the conditional option system that we met in the introduction to BATS. We have not modelled seasonality before so, not surprisingly, this is the first appearance of these seasonal component output options. Since we've talked about them, why not have a look at what they do?

▷ Select **Seas**, then **Both**.

A two frame display is shown giving the seasonal factor estimates in two forms. In the lower frame the actual factor estimates for each time are graphed; the upper frame shows the same values superimposed as perturbations from the underlying estimated level. (Watch out for the vertical scalings in the two frames—they are different.)

▷ Quit from the on-line seasonal estimates graph.

Another similar graph appears—this time for the retrospective estimates. The **Both** option is responsible for this. If we had just wanted to view the on-line or the retrospective estimates we would have chosen the appropriate

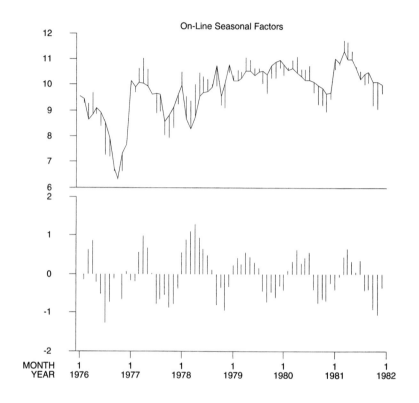

Figure 10.15 On-line seasonal estimates.

alternative menu entry at the last but one step above. The two displays are reproduced in Figures 10.15 and 10.16 respectively.

▷ Quit from the smoothed seasonal estimates graph.

The other seasonal specific option on the **Explore** menu is **Factor**. This lets us look at the development of selected seasonal factor estimates. For example, if we want to see how January sales is seasonally perturbed from the underlying level we take a look at **Factor 1**.

▷ Select **Factor**.
▷ Choose **O Factor 1** and **S Factor 1** (highlight and press **ENTER** for each in turn).
▷ Press **ESC**.

A graph comprising two frames is drawn (Figure 10.17). The right hand

10.11 Extending the Model: Seasonal Patterns

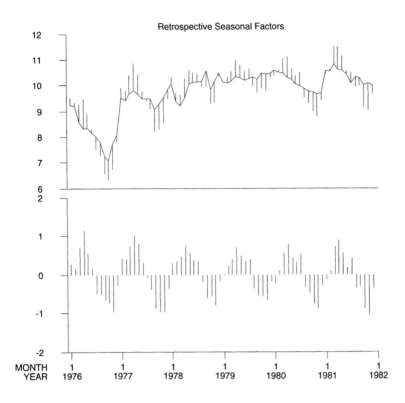

Figure 10.16 Retrospective seasonal estimates.

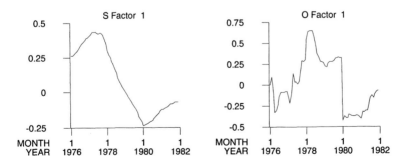

Figure 10.17 Factor 1, on-line and smoothed.

frame graphs the on-line estimate of January's seasonal sales estimate (remember that the O prefix stands for on-line, S for smoothed). The left hand frame graphs the corresponding smoothed estimate. Only the point estimate is shown, these graphs being too small to include uncertainty limits with any clarity—the limits can be seen by viewing factor estimates using the **Time-Plot** or **MPlot** options if you wish.

January's seasonal factor seems to be quite interesting. The most obvious feature is that over the five years spanned by the data there are substantial periods of time when January is a higher than average sales month, and conversely substantial periods of time when January is a lower than average month. In other words, there is evidence of a changing seasonal pattern. So much for static models! Another interesting feature is the way in which the switch from above average to below average takes place. This is best seen from the on-line estimate. There seems to be an abrupt change from a positive estimate to a negative estimate at the beginning of 1980. Such abrupt changes are not reflected in retrospective estimates in the kind of analysis we have done here because the overall fit is made as smooth as possible given the model form and the dynamic variability governed by component discount factors. In common with the previous analyses we did not explicitly set a seasonal component discount factor—or indeed any other—so the default value of 0.95 was used for all times. (The trend and regression default discounts are 0.9 and 0.98 if you recall.) This value is greater than the trend discount reflecting the observation that seasonal patterns are typically more stable than trends, but less than the regression discount. Modelling and analysis of abrupt parameter changes are discussed in detail in the next tutorial.

We can get another view of the contribution of the seasonal component to the sales estimate by looking at time plots of **O Seasonal** and **S Seasonal**. We will just look at the smoothed estimates here.

▷ Quit from the **Factor** plot.
▷ Select **Time-Plot**.
▷ Choose **S Seasonal**, press the **right arrow** key to move to the second column, then the **down arrow** repeatedly to highlight **S Seasonal**; press **Enter**).
▷ Press **ESC**.

The smoothed estimate of the seasonal component (Figure 10.18) seems to show that (i) there is a definite seasonal variation in the sales series that is not reflected in the price variable and (ii) the pattern is quite regular over the six years of the data but certainly not constant. The pattern is also quite smooth and reasonably symmetric which suggests that a parametric form simpler than the (constrained) 12 factor model may be preferred. The next section discusses modelling with restricted seasonal components.

10.11 Extending the Model: Seasonal Patterns

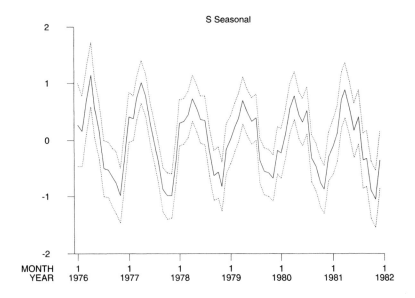

Figure 10.18 Timeplot of smoothed seasonal estimates.

▷ Quit from the graph.
▷ Quit successively to the root.

10.11.2 Restricted Seasonal Patterns

Monthly data exhibiting cyclical behaviour on an annual basis requires 12 individual parameters—so-called seasonal factors—to model that behaviour. Conventionally it assumed that over a full year the sum of the factors is zero—whence they are called seasonal effects; so that individual effects model a monthly seasonal departure from an underlying level. This is an example of the decomposition of complex series behaviour into simple component parts.

The free form seasonal component allows complete freedom in the shape of the pattern exhibited by the seasonal factors. In practice, it is very often found that such great flexibility is not needed. Recall our remark at the end of the analysis in the previous section: we concluded that the regularity of the seasonal pattern over a year was evidence that a simple functional form may be more appropriate than a fully unrestricted model. In this section we show how to define restricted seasonal forms. To begin the study we review a little background theory (see Chapter 3 for a more technical discussion).

The Fourier Representation theorem states that an arbitrary cyclical pattern may be represented exactly by a sum of the cycle's harmonics. Harmonics are simply cosine waves with individual harmonics distinguished by the number of complete cycles they execute for a given period. For example, monthly data with an annual cycle has a seasonal period of length 12. The mathematics of harmonic analysis tells us that there are 6 harmonics for a cycle of length 12, these harmonics having periods 2, 3, 4, 6, and 12. In other words, a sum of 6 cosine waves with cycle lengths as just listed can represent exactly *any* seasonal pattern with period 12. The harmonic with period 12 is called the fundamental harmonic. It completes precisely one cycle over the full 12 time intervals. The second harmonic, with period 6, completes precisely 2 full cycles over that time; and so on for the remaining harmonics.

Many time series in commerce, engineering, and other fields, exhibit seasonal fluctuations that are extremely well modelled by just one or two harmonics—the full set being unnecessary. 'Unnecessary' or 'superfluous' should start alarm bells ringing in your modeller's head. The principle of parsimony—or modelling common sense—tells us that simpler models are preferred to more complex models when explanatory power is not diminished by the absence of excluded terms. When modelling seasonal patterns, if a restricted set of cycle harmonics explains the seasonal variation, we should not include any extra harmonics in the model just to make up the set. Indeed, doing so—using a full harmonic, or unrestricted, seasonal model where the underlying pattern of variation is basically restricted—means that prediction uncertainty will be unnecessarily increased.

To exemplify the prediction problem, consider a yearly seasonal pattern deriving from just the first fundamental harmonic. The single cosine wave has two defining parameters (pitch and phase). Now consider the effect of modelling the seasonal pattern with the first two harmonics. Even if the parameters of the second harmonic are estimated to be zero, they will have some positive—however small—amount of uncertainty associated with their estimation: the variance of their posterior distributions. So that when a forecast is produced using the (overspecified) model, while the point predictions (predictive means) will be approximately the same as for the simpler model, the prediction variance *must* be larger. And quite unnecessarily so! One consequence of excessive forecast variance is a reduced ability to detect breakdown in forecast performance (see Chapter 11).

Technical Aside

Astute readers will have noticed that the statement just made about variances is not the whole story. Simplifying, we have said that the variance of a sum of uncertain quantities is greater than the sum of the variances of the

10.11 Extending the Model: Seasonal Patterns

individual quantities, $V(A+B) > V(A)+V(B)$. The exact relationship is, of course, $V(A+B) = V(A)+V(B)+2\times \text{Cov}(A,B)$. Now, since covariances can be negative (for inversely related quantities) it is quite possible for the variance of a sum to be *less* than the sum of the variances. However, in the situation discussed above, covariances are typically negligible and our simplified assertion stands.

To summarise what we have said: seasonal patterns may be modelled using harmonic analysis; harmonics offer a convenient way of specifying cyclical patterns subject to certain restrictions on the forms of cyclical variation; and when appropriate, such models provide better forecasts than full free-form seasonal models.

BATS, of course, provides for restricted seasonal pattern modelling using harmonics. That's what this section is all about. So, without further ado, let's see how to specify a restricted harmonic seasonal component, and investigate the results from such a model.

▷ From the root, select **Model, Components, Seasonal**.
▷ Select **Restricted-harmonics**.

Two dialogue boxes appear, one for the period of the seasonal cycle—the number of time units over which one complete cycle is defined—and one for harmonic indices. The sales data is recorded monthly and the natural seasonal cycle length is one year, or 12 time intervals. Many economic and commercial series are driven by annual seasonal cycles but there are exceptions, so BATS does not make any assumptions. Our series is not an exception.

▷ Enter 12 for the period.

You may recall from the previous section that the **Free-form** option for selecting a full harmonic model assumes the seasonal period to be the same as the data period. This will typically be the case (as it is here), but there may be occasions when a different period is desired. The period specification in the **Restricted-harmonics** option allows such models to be built (even for full harmonic models since you are free to enter all harmonic indices in the relevant box). We shall include just the fundamental harmonic in the model.

▷ Enter 1 in the 'harmonic indices' box.
▷ Press **ESC** to have the entries loaded.

For comparison with the earlier analysis, use the reference initialisation. We will compare the estimated seasonal component from this model with the estimate from the full harmonic model of the previous section.

▷ Quit to the root.
▷ Select **Fit**, then **Reference**.

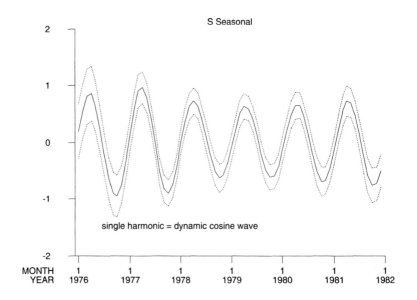

Figure 10.19 Estimated seasonal pattern: single harmonic model.

- ▷ Quit from the on-line graph.
- ▷ Select **Retrospective**.
- ▷ Quit from the retrospective graph.
- ▷ Select **Data, Explore, Time-Plot**.
- ▷ Press the **right arrow** key to move the highlight to the second column.
- ▷ Press the **down arrow** key (repeatedly) to highlight **S Seasonal**.
- ▷ Press **ENTER**, then **ESC**.

Figure 10.19 shows the estimated (retrospectively fitted) seasonal component for the single harmonic model. It is a dynamic cosine wave. Had we set the seasonal component discount factor to 1.0 for a static model, there would be no peak-to-trough variation in the quantification of the cosine. (There would be no phase shift either; but as you can see this analysis shows no evidence of phase shift anyway.) Compare Figure 10.19 with Figure 10.18. You can see that the overall general shape of the seasonal pattern in the free form model is very well captured in the single harmonic model.

We have now seen two seasonal models for the sales data: the free-form model and a single harmonic model. Which model is appropriate? The classical tool used to investigate this question is the periodogram. The same tool may be used here too. You must bear in mind though that the periodogram assumes a stationary series, so the results must be examined

10.11 Extending the Model: Seasonal Patterns 295

carefully. Quite clearly, the analysis in this chapter shows that the sales data is *not* stationary. There is really no substitute for thinking about the real problem and comparing the performance of alternative plausible models. Starting out with an unrestricted model and examining estimated harmonics and comparing predictive performance are the best routes to take when doing this.

10.11.3 Prediction

To round off this tutorial take a final look at prediction.

▷ Quit from the time plot, then successively to the root.
▷ Select **Predict**.
▷ Set the forecast horizon to December 1982 (1982/12); press **ESC**.
▷ Leave the regressors as zero; press **ESC** to quit the table.
▷ Select **Continue** to use the model based prior moments.

(As with the filter operations the prediction calculations for this model take a little longer than for simpler models—you may notice a second or so delay, more for longer forecast periods and models with several harmonics.) The predictions are not constant this time: there is seasonal variation about the constant level, Figure 10.20. But there is no contribution from the regression because we left the price variable set to zero. Figure 10.14 shows predictions for a comparable analysis for the nonseasonal model.

Figure 10.21 shows predictions from the full seasonal model. Comparing these with the predictions from the single harmonic seasonal model illustrates the point we raised earlier about excessive forecast variance from overspecified models.

When you have finished looking at the predictions you can go ahead and terminate the session since we have come to the end of this tutorial.

▷ Quit the prediction graph.
▷ Quit the **Forecast** menu, and successively to the root.
▷ Select **Exit**.
▷ Select **Yes**.

BATS returns you to the operating system (DOS) in the directory from which you began the session.

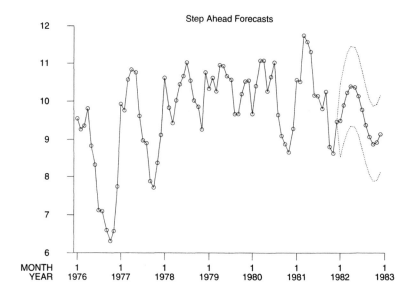

Figure 10.20 Predictions from single harmonic seasonal model.

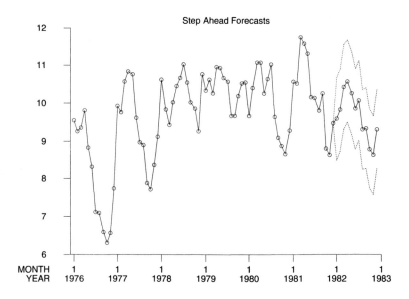

Figure 10.21 Predictions from full seasonal model.

10.11.4 Prior Specification for Seasonal Components

In all of the analyses in this chapter, we have used the reference initialisation for convenience. We also saw the specification of component priors for trend and variance components in the very first prediction section. Specifying priors for a regression component is handled similarly: one specifies a prior mean and standard deviation for the parameter of each regressor included in the model. Now, what about prior specification for seasonal components? Knowing that BATS builds seasonal components from sums of harmonics, does this mean that we have to specify priors on these harmonics? And what about ensuring that the zero sum constraint is complied with?

Don't worry! BATS does the tiresome work for you. Information about seasonal components is *always* communicated through the medium of seasonal effects—deviations (monthly, quarterly, or whatever the timing interval happens to be) from the underlying level. If a yearly seasonal cycle for monthly data is defined, then 12 seasonal effect means and variances define the quantification of the component. You may forget about constraint problems because the program will enforce them. For example, if you specify a set of seasonal effect estimates that sum to $k \neq 0$, then each effect will be adjusted by subtracting an amount $-k/p$ where p is the seasonal period (equivalently, the number of seasonal effects). Effect variances are adjusted by utilising properties of the multivariate normal distribution to ensure the proper singularity of their joint distribution. For restricted harmonic models similar generalised results define properly adjusted effect means and variances. Basically your (arbitrary) pattern of seasonal factors is fitted as best as can be by the functional form of the included harmonics. Mathematical details are given in Chapter 3.

10.12 Summary of Operations

1. Defining a seasonal component
   - ▷ From the root, select **Model, Components, Seasonal**.
   - ▷ Select **Free-form**.
   - ▷ Quit to the root.
2. Sequential reference analysis
   - ▷ From the root, select **Fit**, then **Reference**.
   - ▷ Quit from the forecast graph.
   - ▷ Select **Retrospective**.
   - ▷ Quit from the retrospective graph.
3. Examining results
   - ▷ From the root, select **Data, Explore**.
   - ▷ Select **Seas**, then **Both**.
   - ▷ Quit from the on-line seasonal estimate graph.
   - ▷ Select **Factor**.
   - ▷ Highlight **O Factor 1**, then press **ENTER**.
   - ▷ Highlight **S Factor 1**, then press **ENTER**.
   - ▷ Press **ESC**.
   - ▷ Quit the graph.
   - ▷ Select **Time-Plot**.
   - ▷ Highlight **S Seasonal**, press **ENTER**, then **ESC**.
   - ▷ Quit from the graph, then Quit to the root
4. Defining a restricted pattern seasonal component
   - ▷ From the root, select **Model, Components, Seasonal**.
   - ▷ Select **Restricted-harmonics**.
   - ▷ Enter 12 for the period.
   - ▷ Enter 1 in the 'harmonic indices' box.
   - ▷ Press **ESC** to have the entries loaded.
5. Sequential reference analysis; examining results
   - As for 2 and 3 in this list.
6. Prediction
   - ▷ From the root, select **Predict**.
   - ▷ Set the forecast horizon to 1982/12, press **ESC**.
   - ▷ Leave the regressors set to zero, press **ESC**.
   - ▷ Select **Continue**.
   - ▷ Quit the graph.
7. Concluding the session
   - ▷ From the root, select **Exit**, then **Yes**.

10.13 Items Covered in This Tutorial

The purpose of this introductory modelling tutorial has been to give you a (strong) flavour of the kinds of analyses that you can do with BATS, and to show you how to perform them. We have covered quite a lot of ground so now is a good time to take a break. To finish up, we leave you with a summary of the material discussed.

- Model specification: response, trend, regression, seasonal components.
- Harmonic analysis: free form and restricted harmonic seasonal patterns.
- Setting prior values.
- Prediction: before analysis, following analysis, setting future regressors, comparative 'What if?' predictions.
- Analysis: on-line and retrospective.
- Reference initialisation.
- Data analysis: one step forecasts, on-line estimates, retrospective estimates.
- Examination of results: general graphs and tables, specialised graphs.
- Interpretation of results.
- Setting component discount factors.
- Data transformations, creating new series.

Chapter 11

Tutorial: Advanced Modelling

Chapter 10 introduced the basic component structures which may be combined to build a dynamic linear model with sufficient complexity to model a wide range of univariate time series. We built a complex model comprising a steady trend, a single variable regression, and a seasonal component. We explored prediction, dynamic analysis (on-line forecasting and retrospective filtering), and the effects of altering component discount factors and prior means and variances.

In this chapter we take this modelling and analysis to a higher level of sophistication, introducing forecast monitoring, subjective intervention, and modelling of variance laws.

11.1 Data and Model

The data shown in Figure 11.1 is the number of car drivers killed and seriously injured in Great Britain (source: Civil Service private communication, but the data is publicly available and is featured in Harvey and Durbin, 1986). If you look at this time series for a moment, the general features should be relatively easy to see.

- An overall level trend that seems to be broken into three separate phases with breaks at the beginning of 1974 and the beginning of 1983.
- A pronounced seasonal pattern that seems fairly consistent across the entire series.

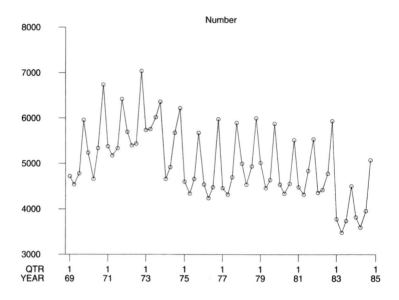

Figure 11.1 Time plot of quarterly injuries.

A first pass at modelling this series is clearly going to involve investigating a model comprising a trend component and a seasonal component. This will be our starting point for the tutorial. We saw in Chapter 10 how to specify both trend and seasonal components so we will just list the required steps here. If you need a reminder of what each of these steps is actually doing, refer back to the relevant sections of Chapter 10. Begin by loading the injury data set QACCIDS.DAT.

▷ From the root select in turn Data, Input-Output, Read, BATS format.
▷ Highlight QACCIDS.DAT then press ENTER.

(We are assuming here that the QACCIDS.DAT data file is stored in the BATS format data directory. Refer to the appendix to Chapter 9 if you need help with file locations.) To specify the model, return to the root then set the response, trend and seasonal components.

▷ Quit to the root.
▷ Select Model.
▷ Press ENTER to select Number for the response.
▷ Select Components.
▷ Select Trend, then Linear.
▷ Select Seasonal, then Free-form.

We have done things a little differently for this model compared to what we did in Chapter 10. For the trend we are using a linear component where before we had a steady component. If you take a look at the first 'segment' of the data in Figure 11.1 an upward underlying trend is quite discernible. (To see the graph on the screen, return to the root, then select **Data, Explore, Time-Plot**. Mark **Number** by pressing **ENTER**, then press **ESC**. Press **Q** to quit the graph as always.) A steady model is too restrictive to capture such development—we need a trend component linear in time. For the seasonal component we have set, as with the example in Chapter 10, an unconstrained pattern. This time however, since the data is quarterly there are just four data points which span a complete yearly cycle. So the required period is four and the necessary harmonic indices are one and two. We could have set these values explicitly by using the **Restricted-harmonics** option, but since the data period is already set to four, the **Free-form** option was less work.

11.2 Preliminary Analysis

Examination of Figure 11.1 shows that within each of the segments 1969–73, 1974–82, and 1983–84, the injury series is quite stable, exhibiting apparently little movement in the trend and cyclical patterns. Furthermore, the cyclical pattern seems to be reasonably constant over all three segments. This stability leads us to expect that component discount factors for a routine analysis can be quite high—certainly higher than the BATS defaults. The only practical fly in this analytical ointment is going to be how we deal with the instabilities or breakpoints which divide the series into the three identified segments.

One possibility is to analyse each segment separately, then compare the three sets of component estimates. The problem with that approach is that we lose information when some components are not affected by the causes underlying the segmentation. In the injury series we have already postulated that the seasonal pattern seems to be consistent across all three identified segments. If it is, we don't want three lots of seasonal component estimates. Fortunately, segmentation—or structural change—is easily modelled within the Bayesian dynamic linear model framework. Even better, the BATS implementation can handle it too!

To begin the analysis we will increase the component discount factors in line with our comments on the stability of this data. (We discussed the alternatives **Constant** and **Individual** in Chapter 10.) Use the reference prior initialisation.

▷ Quit to the **Model** menu. (If you have been viewing the data using **Explore**, then quit the graph, quit to the root, and then select **Model**.)
▷ Select **Discount**.

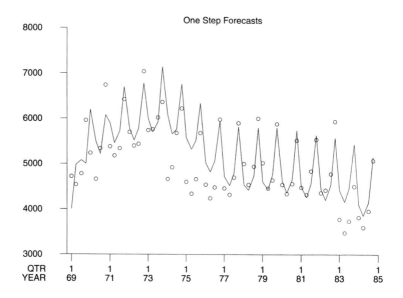

Figure 11.2 One step forecasts.

▷ Select **Constant** and set the trend and seasonal discounts to 0.98. (The down arrow key moves the cursor from box to box; pressing ESC finishes the input.)
▷ Quit to the root.
▷ Select **Fit**, then **Reference**.

Notice from the pattern of one step ahead forecasts (Figure 11.2) how the seasonal pattern is unknown at first and has to be learned about over the course of the first year. This is a direct feature of the reference initialisation which always assumes great uncertainty and zero knowledge of such patterns. Notice also what happens to the forecasts after the downward shifts of the data in 1974 and 1983. The model is quite slow to react to these changes, particularly the first where there is both a drop in level and a reduction in underlying growth. It takes about three years before the forecasts are adequately adjusted. The change in 1983 is less severe and we expect forecast performance to improve more quickly than before (because it does not have 'so far to go'). This does seem to be the case but unfortunately there is insufficient data following the break to confirm this. This slow-to-adjust feature is a direct result of the high discount factors that we set. A suggestion for improving the speed with which the model adjusts to the changes in 1974 and 1983 would seem to be to lower the discounts thereby allowing greater dynamic evolution at each time. But what about

11.2 Preliminary Analysis

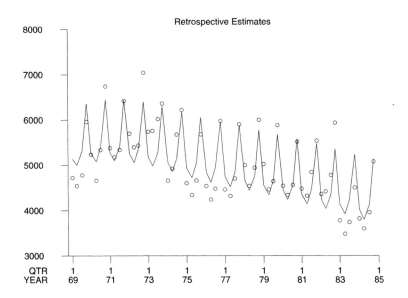

Figure 11.3 Retrospective estimates.

everything we said at the start in justification of high discounts? Is it not true that the stability of which we spoke there *is*, in fact, exhibited? We seem to have two opposing factions in this data.

The adversarial nature of this situation points up the problem with looking for an 'optimal' set of component discount factors that are appropriate for all time. Structural instabilities should be modelled directly, not fudged by 'playing both ends against the middle'. Lowering the routine discount factor will make adaptation to change easier (and therefore faster) but it will also increase forecast uncertainty throughout the entire analysis which is clearly undesirable. Model forecast uncertainty should ideally be reflecting how sure we are about what we 'know' at each time. In the injury series, outside of the changes around the start of 1974 and 1983, matters are indeed quite stable.

Before we consider what may be done about the stability versus structural change conflict let's take a look at the fitted estimates for the 'stable' analysis (Figure 11.3).

▷ Quit from the on-line graph.
▷ Select **Retrospective**.

The picture here is quite interesting. The fitted values have an extremely uniform seasonal pattern superimposed on a slowly undulating trend which has been fitted through the data as smoothly as possible. That is exactly what we should expect from this analysis. The fundamental model form is a cycle composed with a straight line. The model dynamics allow a little movement from these strict descriptions so that we can have a trend that is *locally* linear but which can adjust a little through time. The seasonal pattern may also adjust slowly over time. Sudden large shifts in parameter estimates are not allowed.

11.3 Monitoring Forecast Performance

It is all very well for us to look at the foregoing analysis and pronounce upon its shortcomings. After all, we have the benefit of viewing the data as a whole: we can see the segmentation at 1974 and 1983. Unfortunately, in a live situation we typically do not have the benefit of foresight. However, we do not have to be quite as simplistic as our first analysis assumes. Just because we have specified a model form and a particular quantification based on past experience does not mean that we cannot make changes in response to new information when it becomes available. Indeed, it is precisely this ability to adjust 'as and when' which is a great advantage of the dynamic Bayesian approach.

The first thing we can do is to compare our forecasts with the actual injury figures as the numbers become available. Then, if we notice any 'odd' discrepancies we can investigate matters more closely. And it would be convenient if we could have our forecasting system somehow police itself in this regard. The system 'knows' the forecasts it generates and receives the outcomes when they become available so all that is missing is a rule for deciding which comparisons fall into the category of 'odd discrepancy'. The general theory for such forecast monitoring is described in Chapter 3.

BATS provides three types of forecast watchdog, or monitor. One for specifically detecting upward changes in underlying trend; another for detecting downward changes; and a third for detecting nonspecific discrepancies. Any or all of these monitors may be used to check forecast performance.

11.3.1 Setting a Monitor

We shall make use of the 'downward shift in trend' monitor to illustrate the mechanism.

▷ Quit from the retrospective graph.
▷ Select **Model, Interrupts**, then **Monitor**.

As you can see from the screen display (Figure 11.4) the three monitors are all turned off by default. You can turn a monitor on by moving the block

11.3 Monitoring Forecast Performance

```
            Scale Inflation Monitor:      N

           Level Increase Monitor:        N

           Level Decrease Monitor:        N
```
Figure 11.4 Monitor specification screen display.

cursor to the corresponding dialogue box and changing the **N** to a **Y** (or **y**). Recall that the keyboard arrow keys are used to move among a series of dialogue boxes.

▷ Use the down arrow key to move to the box opposite the label 'Level Decrease Monitor' and press **y**.
▷ Press **ESC** to enter the settings.
▷ Quit successively to the root.

11.3.2 Analysis with Monitoring

We are now ready to repeat the analysis of the injuries series, this time with a check performed on the forecast accuracy.

▷ Select **Fit**, then **Reference**.

The analysis is the same as before! Monitoring for a downward shift in level did not make a difference after all. Why not?

The basic reason is the sensitivity of the monitor. When we compare a forecast with a subsequent outcome we have to decide by how much we will allow the numbers to diverge before we consider the situation as poor forecasting performance. In the present analysis, BATS decided that despite the quite large absolute errors in the forecasts beginning in 1974 they were not large enough to signal a potential problem. You may disagree! The crux of the matter hinges on the relative size of the forecast error, not the absolute size. For a forecast about which we are extremely certain we would want to stop and think again if the outcome differed by a smaller amount than if we had been rather uncertain about the forecast. So the forecast error should be weighed in terms of the forecast uncertainty. And it is precisely this point which explains BATS' failure to signal in the current analysis. The reference initialisation set prior values which had extremely great uncertainty in all components and little idea about component forms. Even by the time 1974 is reached the forecast distribution is still very diffuse so that the absolute forecast error is not large in comparison with the forecast standard deviation: so no signal. You can see how uncertain

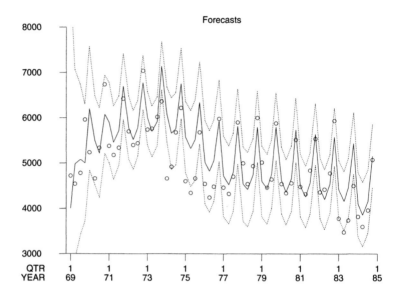

Figure 11.5 One step forecasts.

the forecasts are by quitting the graph and examining the forecasts from the **Explore** menu since those time plots include 90% uncertainty limits (Figure 11.5). Notice how the uncertainty increases as the model reacts to the lower values than expected from the beginning of 1974 on.

▷ Quit from the on-line graph.
▷ From the root select **Data**, **Explore**, then **Time-Plot**.
▷ Highlight **Forecasts**, press **ENTER** then **ESC**.

Bearing the relativity concept in mind, take a look at a table of the forecasts and injury series (Table 11.1).

▷ Quit the graph.
▷ Select **Table**.
▷ Highlight **Number** and **Forecasts** in turn and press **ENTER**; press **ESC**.

You might think that the number of injuries in the first quarter of 1974, 4659, is sufficiently smaller than the 90% lower forecast limit, 5474 (calculated by subtracting the 90% value, 594, from the forecast mean, 6068), that a signal should have been generated. The reason no such signal was produced is that the default setting of the monitor allows for quite substantial forecast inaccuracy. We will see shortly how this sensitivity may be tailored to reflect any required system performance characteristics.

11.3 Monitoring Forecast Performance

Table 11.1 Injuries and forecast series

Time	Number	Forecast	90%
1973/1	5737	5984	583
1973/2	5737	5688	566
1973/3	6008	5916	543
1973/4	6348	7137	524
1974/1	4659	6068	594
1974/2	4911	5642	785
1974/3	5670	5750	801
1974/4	6220	6753	783

Before we adjust the monitor let us repeat the analysis with reduced uncertainty in the starting estimates for the parameters.

▷ Press ESC to quit the table.
▷ Quit to the root and select Fit.
▷ Select Existing-Priors.

The Existing-Priors option is a quick way of reusing the prior from the most recent analysis. Choosing this allows us to (examine and) alter the priors which BATS calculated for us on the previous (reference) run. Resetting priors operates in the same way as setting priors for a prediction (you may recall this from the last tutorial). We will retain the component prior estimates (the parameter means) but reduce the variances about these values. The variances will still be 'large', that is, reflecting substantial uncertainty, but not as large (uncertain) as before.

▷ Select Trend.
▷ Reduce the level standard deviation to 500 and the growth standard deviation to 100, then press ESC.
▷ Select Seasonal, then Exchangeable.
▷ Enter 250 for the seasonal effect standard deviation.

In this context 'exchangeable' signifies that all of the seasonal effect means (quarters one through four) are set to 0 and all have the same uncertainty about this value. This is an assumption of ignorance or lack of specific knowledge about the seasonal pattern, although the variance does indicate the overall peak to trough range considered plausible.

▷ Select Quit, then Variance.
▷ Change the estimate of the observation standard deviation to 100; press ESC.
▷ Quit to begin the on-line analysis.

BATS generates an exception signal for the first quarter of 1974. As we saw above, the number of injuries recorded for this quarter is substantially lower

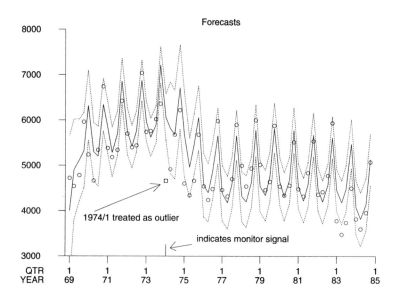

Figure 11.6 One step forecasts with monitoring.

than the forecast and this time the forecast uncertainty is sufficiently small that this difference gives cause for concern. (The forecast here is different from the forecast in the previous analysis because we have used different starting conditions. However, all we did was reduce initial uncertainty; we did not change initial location estimates. The difference in the point forecasts is therefore small because even though the initial uncertainty was reduced it was still large in absolute terms, allowing the model to adjust to the data.) The signal information also informs us that the signal has a run length of one. We will talk about run lengths later so ignore that for the moment.

What to do in response to the signal? BATS offers a choice of three courses of action at this point. We can ignore the signal and proceed with the analysis as if nothing had happened. This is equivalent to the previous analysis where no monitoring was performed. We can opt for the automatic signal response, which allows BATS to decide upon a course of action. Or we can intervene in the analysis ourselves.

The first course of action—actually inaction!—is not very interesting (we've seen it before) so let's take the second and leave it to BATS to sort the problem out.

▷ Press 2.

11.3 Monitoring Forecast Performance

The signal information is removed and the on-line analysis continues to the end of the series. Notice the box drawn around the observation for 1974/1 and the comment in the model summary panel, 'Outlier: 74/1'. These indicate that BATS' response to the monitor signal was to regard the observed number of injuries for 1974/1 as a missing value—which is to say the value is ignored by the analysis. The monitor signal itself is indicated by a line drawn upwards from the time axis. Notice also that the adjustment of the forecasts to the now lower level is rather faster than in the original analysis (compare Figure 11.6 with Figure 11.5). So what is going on here?

11.3.3 Modes of Automatic Signal Handling

BATS has a very simple set of rules to follow when deciding on a course of action to pursue in the event of a monitor break and our decision to choose automatic intervention. In fact there are only two rules. If a signal of run length one occurs the automatic action is to treat the observation that gave rise to the signal as an outlier and to increase component uncertainties before continuing the analysis. If a signal of run length more than one occurs, the rule is increase uncertainties but do not ignore the observation.

When a single observation is very discordant with a forecast, the monitor signals with a run length of one. Such a point could be a single outlying value, a transient or just plain odd observation, or it could be the onset of a fundamental change in the structure of the series. Without any further information about potential problems at such a point there is no way to decide which of these causes apply. The safe course of action is to ignore the value observed—treat it as if it were just an oddity—and to reflect our greater uncertainty by widening the prior standard deviation limits on the component parameters. The forecast for the time following the 'odd' value will now be more uncertain than it would otherwise have been and so more attention will be paid to the actual outcome when updating estimates of component parameters. In other words, the model will adapt more rapidly to subsequent observations because this discrepant value makes us wonder more strongly about the relevance of the past data to potentially changed conditions.

A monitor signal does not always have a run length of one. A signal with run length greater than one is generated when some sort of tendency of forecasts to be consistently different from observations has been detected. The keyword here is consistently. For example, when monitoring for downward level shifts as we are doing here a sequence of low forecasts, none of which individually is sufficiently far from the observed value to be of concern, but which taken together suggest the possibility of change. The change does not need to have been gradual to generate such a signal. A sudden change of a sufficiently small magnitude will typically take several reinforcing forecast errors to alert us to the problem, which explains why interpretation of signals is not easy and best done by someone with detailed background knowledge. It also explains why BATS includes just the two straightforward automatic response rules given above.

We mentioned that uncertainties are increased under the rules for automatic intervention. How? BATS knows about a set of component discount factors specially for use in this situation. These 'exception' discounts are much lower than the values used for routine analysis: the default values are 0.1 for the trend and seasonal components, 0.8 for regressions, and 0.9 for the observation variance. Such small discounts greatly increase the component prior uncertainties over their previous posterior values. Exception discounts, along with monitor sensitivity, may be changed as needed.

11.3.4 Customising the Monitor

Reducing the starting prior uncertainty in the last analysis resulted in forecast variances small enough such that the observation in 1974/1 caused the level decrease monitor to signal. That signal was produced with the monitor at its default sensitivity setting. Let's take a look at how we can adjust monitor sensitivity.

▷ Quit from the forecast graph.
▷ Select Configuration, then Monitor.

Figure 11.7 shows the screen display. There is a lot of information in this display; a full description is given in Chapter 15. For the moment we are only interested in the entries for the level decrease monitor. The default setting for the 'standard deviations' parameter is -3.5. This means that an observation must be at least 3.5 forecast standard deviations less than the forecast mean before the monitor will consider the situation unusual. There is more to the monitoring story than just the forecast error of course—the Bayes factor value is also important. The larger the Bayes factor threshold, the more intolerant is the monitor.

Monitor theory is detailed in Chapter 3. We will concentrate on the practicalities here.

▷ Change the 'Level decrease: std deviations' value to -2.5.
▷ Change the 'Level decrease: Bayes factor threshold' to 0.3.
▷ Press ESC.
▷ Quit to the root.

11.3 Monitoring Forecast Performance

```
Scale shift:            inflation factor:       2.582
                        Bayes factor threshold: 0.300

Level increase:         std. deviations:        3.500
                        Bayes factor threshold: 0.135

Level decrease          std. deviations:        -3.50
                        Bayes factor threshold: 0.135

Cumulative run length limit:                    4

Automatic discounts:
                                       trend    0.1
                                    seasonal    0.1
                                  regression    0.8
                                    variance    0.9
```

Figure 11.7 Monitor configuration screen display.

11.3.5 More Analysis

Before proceeding with the analysis using the more sensitive monitor we have to tidy up some loose ends left over from the previous analysis. If you recall, in that analysis the (automatic) intervention strategy was to treat 1974/1 as an outlier and use the exception discounts in calculating the priors for the subsequent time 1974/2. So far as BATS is concerned these two items are still part of the analysis environment—we want to remove them. The **Reset** option provides the most painless method of doing this 'removal'. We can choose to reset discount factors alone, or intervention/outlier indicators alone, or the whole lot. We want to remove the whole lot to restore the starting conditions of the previous two analyses.

▷ From the root select **Reset**.
▷ Select **All**.
▷ Quit to the root.

We are now ready for the new analysis.

▷ Select **Fit**, then **Existing Priors**, then **Quit**.

The **Existing-Priors** option sets the prior component values to those that were used in the previous analysis, so that the present analysis will be starting from the same basis. A good idea for the comparisons from which we wish to draw later. The monitor signals a break at the fourth quarter of

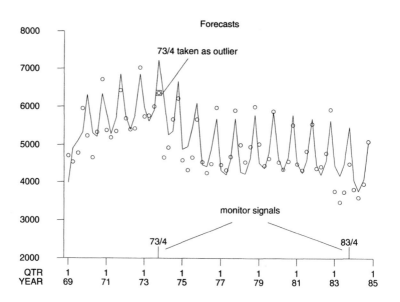

Figure 11.8 One step forecasts with sensitive monitoring.

1973. This is one quarter earlier than in the previous analysis. The more sensitive monitor regards the forecast mean as 'large' where before it was acceptable. As we did before, use the automatic mode of intervention.

▷ Press 2.

The monitor signals a second time, at the final quarter of 1983. You will notice that this second signal reports a run length of four (it was one on the first occasion). So none of the forecasts for 1983 are individually regarded as too discrepant from their outcomes, but taken together, they show up a possible problem. Select the automatic response again.

▷ Press 2.

Notice that, as we discussed earlier, the observation for 1983/4 is *not* treated as an outlier because this time the run length is greater than one.

▷ Quit the forecast graph.

One last point to think about at this time: In the previous analysis a signal was generated for 1974/1. But in the present analysis, with a more stringent set of monitor conditions, no such signal is made. Why not? The reason lies in our response to the signal that *was* generated in this analysis one quarter earlier. If we had taken no action in response to the 1973/4 signal, then 1974/1 would certainly have indicated a break. The analysis would have been identical to the previous case up to this point so

11.3 Monitoring Forecast Performance

a more sensitive monitor would definitely signal where a less sensitive one had done. However we did not opt to do nothing. Instead we treated the 1973/4 observation as 'odd'—we ignored it—but also we used the exception discounts in calculating the component prior distributions for 1974/1. These two actions are the automatic response rule we described above. The prior variances are thereby increased over what they would otherwise have been, and this greater uncertainty feeds directly into the forecast variance. The net effect of this increase in variance is that the large absolute forecast error for 1974/1 is not too large in terms of the forecast uncertainty even by the stricter standards of the newly sensitive monitor; hence no signal.

11.4 Summary of Operations

1. Loading the injury data
   - ▷ From the root select Data, Input-Output, Read, BATS format.
   - ▷ Select data file QACCIDS.DAT.
   - ▷ Quit to the root.
2. Specifying the model
   - ▷ Select Model, Number, Components, Trend, Linear, Seasonal.
   - ▷ Enter 4 and 1 2 for the seasonal period and harmonic indices respectively; press ESC.
   - ▷ Quit to the root.
3. Setting discount factors
   - ▷ From the root select Model, Discount, Constant.
   - ▷ Set the trend and seasonal discounts to 0.98; press ESC.
   - ▷ Quit to the root.
4. Turning on monitoring
   - ▷ From the root select Model, Interrupts, Monitor.
   - ▷ Enter y in the 'Level Decrease Monitor' dialogue box; press ESC.
   - ▷ Quit to the root.
5. Prior specification
   - ▷ From the root select Fit, Existing-Priors, Trend.
   - ▷ Set the level and growth standard deviations to 500, 100 respectively (means unaltered); press ESC.
   - ▷ Select Seasonal, Exchangeable.
   - ▷ Enter 250 for the seasonal factor standard deviation; press ESC.
   - ▷ Select Variance.
   - ▷ Enter 100 for the observation standard deviation; press ESC.
   - ▷ Select Quit to initiate analysis.
   - ▷ Choose automatic monitor signal response, 2.
   - ▷ Quit the forecast graph.
6. Monitor customisation: increasing sensitivity
   - ▷ From the root select Configuration, Monitor.
   - ▷ Change 'Level decrease: std deviations' from -3.5 to -2.5; 'Bayes factor threshold' from 0.135 to 0.3; press ESC.
   - ▷ Quit to the root
7. Reinitialise analysis environment
   - ▷ From the root select Reset, then All, then Quit.
8. More monitored forecasts
   - ▷ Select Fit, Existing-Priors, Quit.
   - ▷ Choose automatic intervention (option 2) at both monitor signals.
   - ▷ Quit the forecast graph.

11.5 Intervention Facilities

Automatic response to a monitor signal is a last-ditch option for when we really have no idea if circumstances have changed or, when we know they have changed but not what the implications will be. For series like the injury data where the elapsed time between successive observations is substantial—three months in this case—there will be plenty of opportunity to investigate. Such investigation will often yield new information relating to the signal. For example, in 1974 the retail price of petrol (gasoline) in the U.K. increased substantially almost overnight as a result of the oil producing cartel OPEC hiking the price of crude oil. (We must remember that OPEC was much more influential in the international oil market in the 1970s than it is in the 1990s—another dynamic model!) One effect of this sudden price rise was a dramatic reduction of the amount of travelling by car in the U.K. Supposing that there is a link between road usage and the number of road traffic injuries, we may conclude that the second quarter of 1974 will show, like the triggering first quarter, a smaller number of injuries than otherwise would have been expected.

Depending upon how much effort is put into this kind of investigative analysis and how far ahead of time we are forecasting we will arrive at a decision to adjust the model forecasts in some way. Let's take a moment to consider how this might be done. Consider our model structure. We have a linear trend component and a seasonal cycle. We ask ourselves, What sort of adjustments are necessary (and justified) to reflect our research findings? It is unlikely that the yearly cycle will change much—although it is conceivable that the peak to trough variation may be reduced. But what about the underlying nonseasonal trend? We determine from our studies that a reduction in level as already evidenced by the 1974/1 'low' value which triggered the monitor will persist and we have doubts about the growth pattern continuing on its positive path—at least in the short term as the economy adjusts. Let's communicate this information to BATS.

First we must reset the analysis environment once more.

▷ From the root select **Reset**, then **All**, then **Quit**.

Initialise the analysis with the priors from the preceding analysis.

▷ Select **Fit**, then **Existing-Prior**, then **Quit**.

The monitor signals as expected in the final quarter of 1973. (Up to this point the analysis is identical to that in the previous section.) This time we are going to control the intervention directly rather than let BATS do it on our behalf.

▷ Enter **3** for user intervention.

```
BATS 2.10              SELECT              Intervention
Available memory:
|View-Priors| Edit-Priors Interrupts Predict Abort Return
Examine prior moments for 74/1
```

Figure 11.9 Intervention menu.

BATS asks us if we want to discard the current observation. The automatic intervention rule that was used in the previous analysis did just that. We do not want to discard the observed value this time; we believe it will not turn out to be a single oddity, but that it contains important post-price-hike information. So we retain it.

▷ Enter **N** (or n) to retain the observed value for 1973/4.

The screen clears the forecast graph and displays the intervention menu, Figure 11.9.

As you can see from the option help bar we will be making interventions for 1974/1. Recall that the monitor signalled at 1973/4 so we have already seen the outcome for that time. We cannot change the past and (in this context) are not interested in altering past analysis. We do want to change how we proceed from now on. The **View-Priors** option allows us to examine the current model based component priors for 1974/1 without changing them. The **Edit-Priors** option allows us to change selected priors if we wish. Before doing so, digress for a moment to look at some other intervention facilities available here.

Abort is shorthand for saying that we do not want to continue with the present analysis—it will be erased. **Predict** is familiar from Chapter 10, except that now we can calculate predictions from a starting point partway through a series. Let's take a look at what the model is currently predicting for the 'next' two years.

▷ Select **Predict**.
▷ Set the forecast horizon to **1975/4** to give two years of forecasts from the current position; press **ESC**.
▷ Select **Continue** to calculate the forecasts.

The forecast graph (Figure 11.10) is quite interesting. You can see that the pattern of predictions for 1974 and 1975 maintains the underlying positive growth evident in the injury series for 1969–73, and the predicted seasonal pattern is also clear. The forecasts are, as expected, too high. Now repeat

11.5 Intervention Facilities

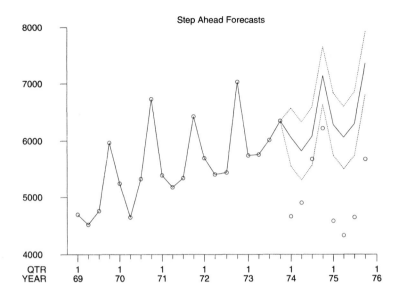

Figure 11.10 Predictions at quarter one, 1974.

the prediction procedure but this time reset the prior estimate for the level to reflect our thinking about the likely effects of the petrol price increase, namely a drop in the level and cessation of underlying growth but with rather greater uncertainty.

▷ Quit from the graph and the forecasting menu.
▷ Select **Predict**.
▷ Enter the forecast horizon as **1975/4**; press **ESC**.
▷ Select **Edit-Priors**, then **Trend**.
▷ Change the level to **5000 500** and growth to **0 25**; press **ESC**.
▷ Select **Quit** to complete prior changes.
▷ Select **Continue** to see the predictions.

Notice that we have not changed the current prior values for either the seasonal component or the observation variance. Our investigations lead us to conclude that any impact the petrol price rise will have on injuries will be on the underlying level alone, so the trend estimates are all that we change. The point forecasts with this modified trend prior are much improved over those from the pure model based prior but the associated uncertainty is also much greater (Figure 11.11).

Would that we could always do this well! We have of course cheated here by looking ahead at the forthcoming data before making our predictions. This is fine for post-hoc analysis and illustration, but in practice we operate

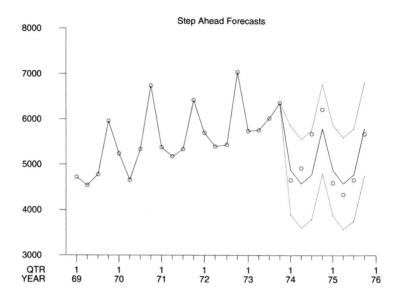

Figure 11.11 Post-intervention predictions at quarter one, 1974.

without such foresight. The points we want to illustrate are the component-wise changes that are possible with these models and how such changes are achieved in BATS. Now let's go ahead with the intervention analysis.

▷ Quit the graph and the forecast menu.

The changes we have just made to the trend component prior were solely for the prediction analysis. This does not alter the prior for the on-line analysis. (So at interrupt times like this we are free to make whatever exploratory examinations of 'What if' scenarios that we please, without affecting the subsequent continuation of the on-line analysis.) We make changes for the on-line analysis separately. (In this case we use the same trend prior that we entered for the predictions a moment ago.)

▷ Select **Edit-Priors** then **Trend**.
▷ Change the level to mean **5000** standard deviation **500**, and growth to mean **0** standard deviation **25**; press **ESC**.
▷ Select **Quit** to finish editing priors.
▷ Select **Return** to continue with the on-line analysis.

The monitor triggers a second time at 1983/4. We will intervene again, retaining the 1983/1 observation and this time changing the trend prior to mean **4000** and standard deviation **150**, and the growth prior to mean **0** and standard deviation **25**. This is a reduction in level and an increase

11.5 Intervention Facilities

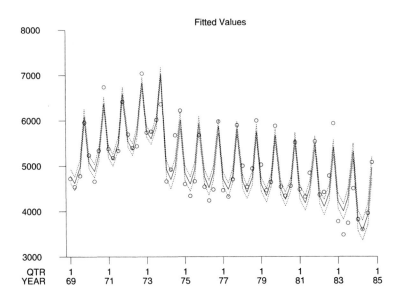

Figure 11.12 Fitted values from intervention analysis.

in uncertainty about the overall trend (level plus growth). Once again attention is focussed solely on the trend, the other components believed to be unaffected by the market changes.

▷ Enter 3 for user intervention.
▷ Enter N to the 'disregard observation' query.
▷ Select Edit-Priors then Trend.
▷ Change the level to 4000 150 and growth to 0 25; press ESC.
▷ Select Quit to finish editing priors.
▷ Select Return to continue with the on-line analysis.

When the on-line analysis is complete run the retrospective analysis immediately, then we will look at both sets of results together.

▷ Quit the on-line graph.
▷ Select Retrospective.
▷ Quit the retrospective graph.
▷ Select Data then Explore.

Figure 11.12 shows the fitted values from this intervention analysis.

▷ Select Time-Plot.
▷ Highlight Fitted Values, press ENTER then ESC.

Comparing this figure with Figure 11.3 the present fit is very good indeed—capturing the trend changes very well, albeit rather late. The smoothed

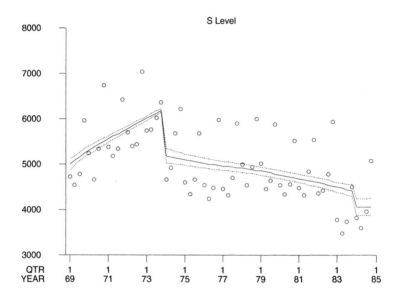

Figure 11.13 Fitted trend from intervention analysis.

level in Figure 11.13 shows the estimated level changes very clearly.
- ▷ Quit the fitted values graph.
- ▷ Select **Time-Plot**.
- ▷ Highlight **S Level**, press **ENTER** then **ESC**.

The downward sloping trend through 1974 to 1984 is something of an artefact of our analysis here. The adjustment we made in 1984/1 followed a monitor break which had a run length of four—signalling that the onset of change was probably back in 1983. However, since the analysis did not catch this change until 1984 the retrospective smoothing assumes stable conditions for 1974/1 to 1983/4. Thus, the downward shift in level in 1983/1 pulls the estimate of the trend down at that time resulting in the overall downward sloping trend. If we were truly analysing these data to identify and estimate changes (up to now we have been simulating live on line forecasting), we would now go back and specify an intervention at the beginning of 1983. The next section illustrates how such forward intervention analysis is performed.
- ▷ Quit the graph.
- ▷ Quit to the root.

At this point you might like to perform another intervention analysis. Suppose that we do not have such good information about the likely impact

11.5 Intervention Facilities

of the oil price rises in 1973 on the trend of road injuries, but that we do believe only the trend is likely to be affected. An appropriate intervention then is to simply increase the trend component uncertainties by some suitable amount. This is a useful analysis to compare with the fully automatic intervention analysis with which we began this study because that analysis increases the uncertainty of *all* model components (using the exception discount factors). We leave this experimentation to you.

11.6 Summary of Operations

This summary covers user intervention in response to a monitor break, and step ahead prediction at an intervention point.

1. Reinitialise analysis environment
   ▷ From the root select **Reset**, then **All**, then **Quit**.
2. Initiate analysis
   ▷ From the root select **Fit, Existing-Priors, Quit**.
   ▷ At the monitor signal select option 3 (user intervention), then respond **N** to 'disregard observation' query.
3. Signal 1973/4: Prediction
   ▷ Select **Predict**.
   ▷ Set forecast horizon to 1975/4; press ESC.
   ▷ Select **Continue**.
   ▷ Quit the graph; quit the **Forecasting** menu.
   ▷ Select **Predict**.
   ▷ Set forecast horizon to 1975/4; press ESC.
   ▷ Select **Edit-Priors, Trend**.
   ▷ Set the level prior to 5000 500, growth prior to 0 25; press ESC.
   ▷ Quit the prior menu, then select **Continue**.
   ▷ Quit the graph; quit the **Forecasting** menu.
4. Signal 1973/4: Subjective intervention
   ▷ Select **Edit-Priors, Trend**.
   ▷ Set the level prior to 5000 500, growth prior to 0 25; press ESC.
   ▷ Quit the prior menu, then select **Return**.
5. Signal 1983/4: Subjective intervention
   ▷ At the monitor signal select option 3.
   ▷ Select **Edit-Priors, Trend**.
   ▷ Set the level prior to 4000 150, growth prior to 0 25; press ESC.
   ▷ Quit the prior menu, then select **Return**.
6. Retrospective analysis
   ▷ Quit the forecast graph.
   ▷ Select **Retrospective**.
   ▷ Quit the retrospective graph.
7. Examination of results
   ▷ From the root select **Data, Explore, Time-Plot**.
   ▷ Select required series by highlighting and pressing **ENTER**; press ESC with all selections made.
   ▷ Quit to the root.

11.7 Forward Intervention

We suggested a moment ago that having identified change points in the injury series in 1973/4 and 1983/1 we should make our interventions at these times. A retrospective analysis will then give accurate estimates of the level and growth changes.

Forward intervention is also useful in forecasting (as opposed to time series analysis) when one knows in advance about a possible change. Obvious commercial scenarios are advertising and other promotional activities for example. In the context of the injury data back in 1982–83 we may also have had advance knowledge of a possible change. Indeed this was certainly the case. New parliamentary legislation passed in 1982 became law on January 31, 1983. The new law made it mandatory for all front seat occupants of motor vehicles (with some minor exceptions) to wear a seat belt at all times during the vehicle's motion.

Let us see how to perform forward intervention analysis with BATS. To begin, clear all existing marked outlying points, exception discounts, and intervention indicators.

▷ From the root select **Reset**, then **All**, then **Quit**.

Now specify the identified change points, 1973/4 and 1983/1, as forward intervention times for a new analysis.

▷ Select **Model**, **Interrupts**, then **Outlier/Intervention**.
▷ Use **Page Down/Up** and the cursor arrow keys to highlight the 1973/4 entry in the intervention column, type **1** and press **ENTER**.
▷ Similarly change the 1983/1 intervention entry from 0 to 1.
▷ Press **ESC** to set the changes.
▷ Quit to the root.
▷ Select **Fit**, then **Existing Priors**, then **Quit**.

At 1973/4 the analysis halts and the display indicates that a forward intervention point is set. Notice that the observed value for 1973/4 is not yet included on the graph. The intervention is *prior* to 1973/4 (the time specified) so changes made are 'before seeing' the outturn. This is in contrast to the monitor intervention we saw above. There we intervened *after* the 1973/4 observation was examined and declared to be at odds with the model forecast. While you are looking at the graph you can see that the preset intervention points are recorded in the summary panel. Go ahead and intervene.

▷ Press **Y** (or **y**) in response to the intervention query.

The intervention menu is precisely the same as we saw when intervening following a monitor break in the previous analysis. Without further ado

make the same changes to the trend as we did before: reduce the level estimate, set growth to zero, and increase level and growth uncertainties.
- ▷ Select **Edit-Priors**, then **Trend**.
- ▷ Change the level to mean **5000** standard deviation **500** and growth to mean **0** standard deviation **25**; press **ESC**.
- ▷ Select **Quit** to finish editing priors.
- ▷ Select **Return** to continue with the on-line analysis.

The analysis halts again at 1983/1. We will intervene on the level exactly as we did for 1983/4 in the last analysis. Once again this involves reducing the current level estimate and setting the growth to zero while increasing uncertainty on both. The remaining component quantifications are not changed.
- ▷ Press **Y** (or **y**) in response to the intervention query.
- ▷ Select **Edit-Priors**, then **Trend**.
- ▷ Change the level to mean **4000** standard deviation **150** and growth to mean **0** standard deviation **25**; press **ESC**.
- ▷ Select **Quit** to finish editing priors.
- ▷ Select **Return** to continue with the on-line analysis.

Forecasting is not our goal with this analysis; we want to investigate the fitted model estimates following from the identified change points. So quit immediately the on-line analysis is complete and set the retrospective analysis in motion.
- ▷ Quit from the forecast graph.
- ▷ Select **Retrospective**.

Quit from the retrospective graph as soon as it is complete and go directly to the explore menu to examine the results, beginning with the estimated trend elements: level and growth.
- ▷ Quit from the retrospective graph.
- ▷ Select **Data**, **Explore**, then **Time-Plot**.
- ▷ Highlight **S Level**, then press **ENTER**.
- ▷ Highlight **S Growth**, then press **ENTER**; press **ESC**.

Evidently there is good reason to believe that small negative growth—a real decline—in the overall level of injuries follows the sharp 1973 downward shift (Figure 11.14). In 1983, however, the trend appears to revert to a regime of positive growth after the sudden fall off, although there is considerable uncertainty about this. These findings are confirmed by the fitted growth estimates (Figure 11.15).
- ▷ Quit the trend graph.

If you take a look at the fitted seasonal pattern (Figure 11.16) and the overall fitted values (Figure 11.17) for this analysis, there can be little

11.7 Forward Intervention

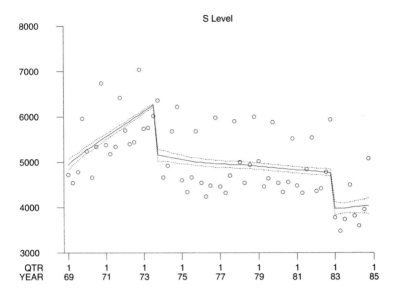

Figure 11.14 Estimated level for forward intervention analysis.

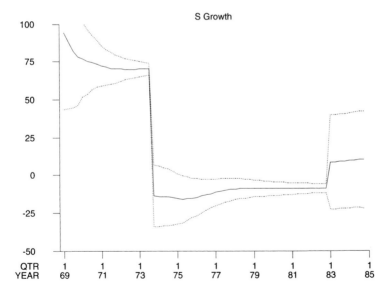

Figure 11.15 Estimated growth for forward intervention analysis.

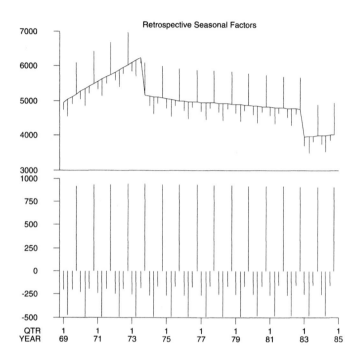

Figure 11.16 Estimated seasonal pattern for forward intervention analysis.

doubt as to the stability of the cyclical variation across the trend changes. Examine the seasonal estimates in the two frame display.

▷ Quit the growth graph.
▷ Select **Seas**, then **Retrospective**.

Examine the fitted values in the familiar time plot.

▷ Quit the seasonal graph.
▷ Select **Time-Plot**, highlight **Fitted Values**; press ESC.
▷ Quit the fitted values graph.

You might like to compare Figures 11.14 and 11.17 from this forward intervention analysis with the corresponding Figures 11.13 and 11.12 from the retroactive intervention analysis of the previous section.

11.7 Forward Intervention

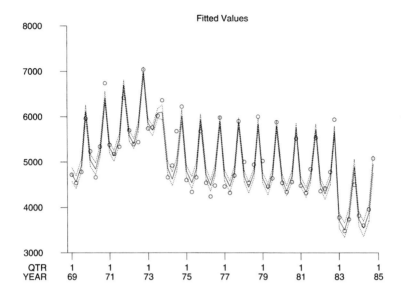

Figure 11.17 Estimates for forward intervention analysis.

11.7.1 Taking the Analysis Further

As we said in Chapter 9 these tutorial analyses are not definitive. They have been designed to demonstrate dynamic linear modelling and the range of features available in the BATS implementation. You have probably already noticed that our latest conclusions about the injury data have glossed over some interesting unexplained features. In particular, despite our claim of structural stability of the seasonal pattern across both changes in level, there is certainly some evidence that this may not be the true story for 1974. There may be more to say about the underlying trend over the 1974–83 period too. Moreover, our monitoring analysis was restricted to using just one of the available monitors—what happens if all three are used?

11.8 Summary of Operations

This summary covers setting times for interrupting an analysis, and making subjective interventions anticipating structural change.

1. Reinitialise analysis environment
   ▷ From the root select **Reset**.
2. Set forward intervention times
   ▷ From the root select **Model**, **Interrupts**, **Outlier/Intervention**.
   ▷ Change the 1973/4 intervention entry (column one) to 1; press **ENTER**.
   ▷ Change the 1983/1 intervention entry (column one) to 1; press **ENTER**.
   ▷ Quit to the root.
3. Initiate analysis
   ▷ From the root select **Fit**, then **Existing Priors**, then **Quit**.
   ▷ Respond **Y** to the intervention query for 1973/4.
4. Forward intervention for 1973/4
   ▷ Select **Edit-Priors**, then **Trend**.
   ▷ Change the level to **5000 500** and growth to **0 25**; press **ESC**.
   ▷ Select **Quit** to finish editing priors.
   ▷ Select **Return** to continue with the on-line analysis.
5. Forward intervention for 1983/1
   ▷ Select **Edit-Priors**, then **Trend**.
   ▷ Change the level to **4000 150** and growth to **0 25**; press **ESC**.
   ▷ Select **Quit** to finish editing priors.
   ▷ Select **Return** to continue with the on-line analysis.
6. Retrospective analysis
   ▷ Quit the on-line analysis graph.
   ▷ Select **Retrospective**.
7. Examination of results
   ▷ Quit the retrospective graph.
   ▷ Select **Data**, then **Explore**.
   ▷ For single time plots select **Time-Plot**, highlight required series and press **ESC**; quit from each graph in turn.
   ▷ Select **Seas**, then **Retrospective** for the two frame seasonal component display.
   ▷ Quit to the root.

11.9 Putting It All Together

One characteristic of the analyses that we have presented in this chapter is the separate treatment of forecast monitoring and prospective intervention. In practice it is more reasonable to be monitoring forecast performance at all times, especially when subjective interventions are made. Such interventions are highly judgemental and one should always exercise great care when evaluating resultant forecasts. Monitoring and feed forward intervention go as much hand in hand as do monitoring and feedback intervention.

As our final excursion in this chapter we bring together some of the issues raised at the end of the previous section. We repeat the most recent forward intervention analysis but now include forecast monitoring with the full set of monitors that the program has available. This analysis will provide some insight into the questions raised at the end of the previous section.

Following the analysis just completed almost everything is already in place: we need only to turn on the forecast monitors.

▷ From the root select **Model, Interrupts, Monitor**.
▷ Enter **Y** (or **y**) in all three of the dialogue boxes to turn on all of the forecast monitors.
▷ Press **ESC**.
▷ Quit to the root.

You may recall that we have not adjusted the monitor sensitivity since we sensitised it earlier in the chapter. The analysis here will therefore be using the default level increase and variance inflation monitors, plus a more than default sensitive level decrease monitor.

Note that it really would not make any difference to the analysis if we were to decrease the sensitivity of the level decrease monitor because we already know what the trigger points are, and they are already preempted here by the forward interventions. So we could in fact remove this monitor from the analysis, but it doesn't hurt to leave it. You may want to experiment later with alternative sensitivity settings for each of the monitors. (Increasing the sensitivity of the level decrease monitor *will* make a difference—at least potentially—can you see why?)

▷ From the root select **Fit**.
▷ Select **Existing-Priors**.
▷ Select **Quit** to begin the on-line analysis.

The monitor triggers at the first quarter of 1970 with a run length of one: the forecast is quite high in comparison with the observed value. Notice that it is the variance inflation monitor that must have signalled here. We know that it is not the level decrease monitor signalling because that monitor did not signal here in a previous analysis when it was the only monitor in place. And it certainly cannot be the level increase monitor signalling because the forecast is greater than the observation.

How should we respond to this signal? Given that we were very uninformative with our prior specification and only four observations have been processed to produce this forecast, the model really has not had opportunity to learn very much yet. We therefore do not want to discard the observation as would happen if we chose the automatic response. The triggering observation contains a lot of information compared to our current state of knowledge. We could intervene and increase component prior uncertainties or make some other subjective intervention, but we just do not have any justification for so doing at this time. We elect, therefore, to ignore this signal and continue with the analysis. This option is always available in practice. Remember: a monitor signal is only one indication of something *potentially* troublesome—even a properly behaved system will occasionally exhibit an observation that will cause a given monitor to trigger, even when nothing untoward is actually occurring. As with other aspects of intelligent monitoring one must not make knee-jerk responses to events; rather, one should think about the possibilities and make a rational decision on a suitable course of action.

▷ Enter 1 for no signal response.

The analysis continues until the forward intervention time set at the fourth quarter of 1973. Make the same intervention as in the previous analysis: reduce the level estimate, set the growth estimate to zero, and increase both uncertainties.

▷ Press **Y** (or **y**) to request intervention.
▷ Select **Edit-Priors**, then **Trend**.
▷ Enter **5000** for the level mean, and **500** for its standard deviation.
▷ Enter **0** for the growth mean, and **25** for its standard deviation.
▷ Press **ESC** to set these trend prior values.
▷ Select **Quit**.
▷ Select **Return** to continue the analysis.

The monitor triggers for the second time at quarter three, 1974. Once again the signal run length is one, indicating a single discordant observation. The forecast is quite low in comparison with the observation. This monitor break is one we have not seen before, not surprisingly since we have not previously been interested in negative forecast errors. Let's just go ahead with the automatic system adjustment: ignore the observation and increase component prior uncertainties.

▷ Enter 2 for automatic signal response.

A third signal—hitherto unseen as with the previous signal—occurs at the first quarter of 1978, again with a run length of one, and again the forecast

11.9 Putting It All Together

is low. Make the automatic adjustment once more.
▷ Enter 2 for automatic signal response.

When the analysis halts prior to the beginning of 1983 make the same intervention as in the previous analysis. The level estimate is decreased and the growth is set to zero, both uncertainties being increased.

▷ Press Y (or y) to select intervention.
▷ Select Edit-Priors, then Trend.
▷ Enter 4000 for the level mean, 150 for its standard deviation.
▷ Enter 0 for the growth mean, 25 for its standard deviation.
▷ Press ESC to set these trend prior values.
▷ Select Quit.
▷ Select Return to continue with the on-line analysis.

When the on-line analysis is complete, quit immediately from the graph and start the retrospective analysis. Our comments will be restricted to an examination of fitted estimates.

▷ Press Q to quit the on-line graph.
▷ Select Retrospective.
▷ Press Q to quit the retrospective graph.
▷ Select Data, Explore, Time-Plot.
▷ Highlight S Trend, S Seasonal, Fitted values, and press ENTER for each in turn.
▷ Press ESC to begin the graph displays.

Figure 11.18 shows the retrospective estimated trend for the present analysis. Comparing this with Figure 11.14 for the intervention-only analysis in the previous section an interesting contrast can be seen. A distinctly lower overall level for the four years beginning with 1974 is indicated by our latest analysis and the corresponding uncertainties are larger. These features result from the increased adaptation made possible by the automatic monitor responses to the signals at 1974/3 and 1978/1.

Figures 11.19 and 11.16 may be similarly compared to examine differences in the inference on the seasonal pattern. As with the trend, there are some contrasts to be drawn. There is evidence of two effects that have not been apparent in our earlier analyses. The seasonal peak seems to be somewhat smaller in magnitude in 1978 and after than it is before. There is also some evidence that up to 1974 the first quarter seasonal factor is smaller than the third quarter, but after 1974 this pattern is effectively reversed. While this is potentially interesting we should point out that the differences in the two quarterly factors are not very large.

The retrospective fitted estimates are shown in Figure 11.20 and may be compared with those in Figure 11.17. The results of the 1974/3 and 1978/1 interventions just described for their componentwise effects can be seen

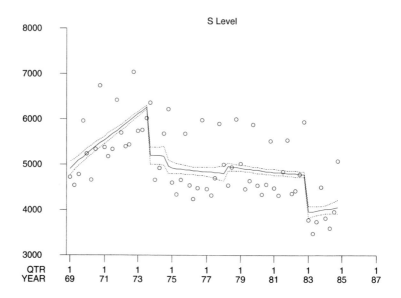

Figure 11.18 Estimated trend.

combined in these figures. Uncertainty limits around both interventions are naturally larger than for the previous analysis, but the evidence of extra adaptation is also apparent. The estimates for 1975–77 in Figure 11.19 clearly provide an improved fit over those in Figure 11.17—a consequence of the combination of the lower level estimates and the higher fourth quarter seasonal factor.

On the basis of the evidence as we have presented it here, it would certainly seem that an excellent case can be made for our most recent analysis as being superior to any previous analysis in this chapter. But it would be premature to stake such a claim just yet. We have only been examining retrospective fit in detail—we have not paid any attention to forecast performance directly, although the monitoring analysis does this indirectly of course. More significantly here, we need to critically examine the interesting effects that we seem to have found: The level increase in 1978, and the already described seasonal pattern changes. None of these changes are very large in magnitude and they could very easily result from general increased volatility in the series over 1973–78 rather than from one or two sharp changes as presented. Certainly the number of injuries in 1973/4 seems quite high, but the trend changes will not trigger the monitors (as set for this analysis) if the trend discount factor is lowered to, say, 0.9 for the

11.9 Putting It All Together

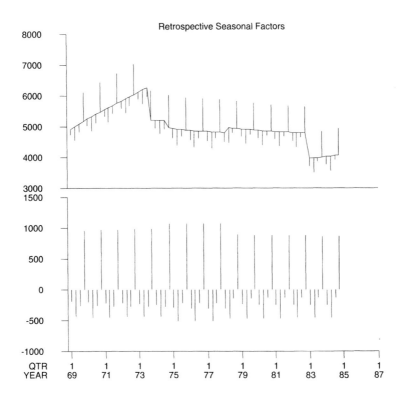

Figure 11.19 Estimated seasonal pattern.

1973–78 period. You may like to verify this claim for yourself. Investigate other discount settings too, both for the trend and seasonal components.

Further exploration of these issues requires more of the same kind of operations we have been doing in this chapter and more background analysis of the features discovered in the data. You should now be in a position to continue the exploration for yourself.

To end the session exit from the program.

▷ Quit to the root.
▷ Select **Exit**, then **Yes**.

BATS returns control to the operating system in the directory from which the session was begun.

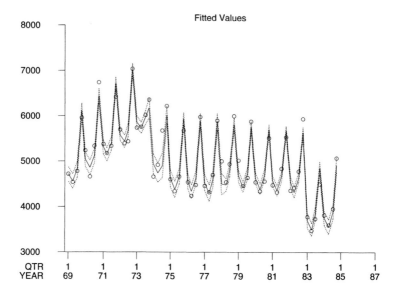

Figure 11.20 Estimated fitted values.

11.10 Summary of Operations

The starting point for this analysis is assumed to be a continuation of the last analysis of the previous section. To initialise the analysis from scratch, define a growth and full seasonal model with trend and seasonal discount factors set to 0.98. Set forward interventions for 1973/4 and 1983/1. Then proceed as listed here.

1. Turn on all monitors
   ▷ From the root select **Model, Interrupts, Monitor**.
   ▷ Enter **Y** (or **y**) in all three of the dialogue boxes to turn on all of the forecast monitors.
   ▷ Press **ESC**.
   ▷ Quit to the root.
2. On-line analysis.
   ▷ From the root select **Fit**.
   ▷ Select **Existing-Priors**.
   ▷ Select **Quit** to begin the on-line analysis.
2.1 Signal response at 1970/1.
   ▷ Enter 1 for no signal response.
2.2 Forward intervention at 1973/4.
   ▷ Press **Y** (or **y**) to request intervention.
   ▷ Select **Edit-Priors**, then **Trend**.
   ▷ Enter **5000** for the level mean, and **500** for its standard deviation.
   ▷ Enter **0** for the growth mean, and **25** for its standard deviation.
   ▷ Press **ESC** to set these trend prior values.
   ▷ Select **Quit**.
   ▷ Select **Return** to continue the analysis.
2.3 Signal response at 1974/3 and 1978/1
   ▷ Enter 2 for automatic signal response.
2.4 Forward intervention at 1983/1.
   ▷ Press **Y** (or **y**) to select intervention.
   ▷ Select **Edit-Priors**, then **Trend**.
   ▷ Enter **4000** for the level mean, **150** for its standard deviation.
   ▷ Enter **0** for the growth mean, **25** for its standard deviation.
   ▷ Press **ESC** to set these trend prior values.
   ▷ Select **Quit**.
   ▷ Select **Return** to continue with the on-line analysis.
3. Retrospective analysis.
   ▷ Press **Q** to quit the on-line graph.
   ▷ Select **Retrospective**.
   ▷ Press **Q** to quit the retrospective graph.
   ▷ Select **Data, Explore, Time-Plot**. Examine graphs as required.

11.11 Items Covered in This Tutorial

In this tutorial we have used the interactive modelling and analysis features available in BATS. We have seen how to analyse time series affected by structural changes and the general component by component adjustment facilities provided in the program.

- Reusing priors from the previous analysis.
- Component prior specification; exchangeable seasonal priors.
- Forecast monitoring: selection of monitor type; monitor configuration.
- Responding to monitor signals: three courses of action.
- Setting forward intervention times.
- 'What if' forecasting at interventions.
- Subjective intervention on component quantifications.
- Forecasting, monitoring, intervention: applying the principle of management by exception.

11.12 Digression: On the Identification of Change

The injury series that we have used in this tutorial was chosen partly because the structural changes exhibited are apparent on a careful visual examination. One can do a very good job of trend estimation—more importantly, estimation of the changes in the trend at the breakpoints—by hand and eye. Sophisticated modelling is hardly necessary for that. One could even go a step further with this and detrend the series 'by hand' and conclude that the seasonal pattern is largely unaffected by the trend changes. It is doubtful anyone could convincingly argue against such a simple analysis in this case.

This situation is not typical of many economic and commercial time series, which often tend to be rather 'noisier' than the injury data. For such noisy series it is very difficult—usually impossible—to determine the nature of any structural changes by visual examination alone. It is very difficult even to decide if there *is* a structural change as opposed to inherent smooth dynamic variation. The applications studied in Chapters 4 and 5 demonstrate this point very well. The industrial sales series graphed in Figure 11.21 is another case in point. Like the injury data the industrial series is quarterly and displays cyclic variability over the year compounded with a moving underlying base level. But how should one characterise this movement in the trend? And what about the seasonal pattern—can we 'see' how stable this is?

One can readily draw a smoothly changing locally linear trend through this data. Equally one could break the series around the first quarters of 1979

11.12 Digression: On the Identification of Change

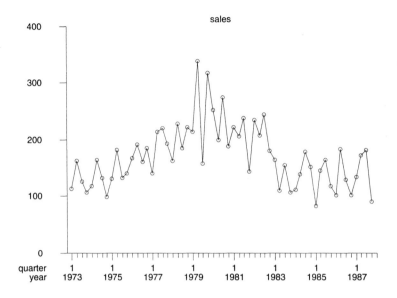

Figure 11.21 Industrial sales series.

and 1983 and draw three quite different locally (almost constant) linear trends. Which model is appropriate? This is a hard question to answer. Visual inspection of the series is simply not enough in this case, unlike that of the injury data. Formal modelling and analysis are necessary.

We are not making any assumptions in this inspection analysis about the state of knowledge of the sales series on the part of the investigator. Even if the investigator has greatly detailed knowledge of the market conditions, a decision about the nature of the trend in the series cannot be made without modelling. Why not? 'Expert' knowledge will enable the investigator to criticise modelling and analysis conclusions from an informed position. For example, there may have been good reasons why a change might have occurred in 1979 or 1983. Or there may not. But such knowledge is not sufficient to convince other people that there actually was a change without sound demonstrative analysis because the conclusions do not leap out of the data as they do with the injury series. And being unobvious, there is room for argument.

We hope that we have piqued your interest in the issues raised by this brief aside. But we want to tantalise you further by not telling you 'the answer' to the questions that we have just outlined about the sales series! The data set is included on the program diskette so you can perform your own analysis and come to your own conclusions.

One last point. This discussion has been couched in terms of retrospectively examining a time series. Imagine how much more difficult the decision problems would be in a live forecasting situation. Chapter 4 demonstrates the difficulties and how they are tackled.

11.13 References

Harvey, A.C. and Durbin, J. (1986). The effects of seat belt legislation on British road casualties: a case study in structural time series modelling (with discussion). *J. Roy. Statist. Soc.* (series A), **149**, 187–227.

Chapter 12

Tutorial:
Modelling with Incomplete Data

When we speak of a complete data set we mean a data set whose constituent series have known values at every timing point defined over a given interval. For example, if we have a data set comprising monthly series for sales and price from January 1985 to August 1991, then if the monthly sales and price values are recorded for all of these months, the data set is complete.

The data sets that we have utilised in previous chapters are complete in this sense. It is now fairly clear what we mean by an incomplete data set: a data set in which there is at least one component series for which one or more of the 'observations' are unknown. In the context of the foregoing example, this could be a missing sales value for September 1987 say. Such 'missing values' may be missing for any number of reasons; indeed a value may not be missing at all, merely regarded as rather unreliable. The reasons do not concern us directly at this point, but whatever they may be, the fact is the values are not known and our modelling and analysis must be capable of dealing with the situation.

Data series with values unavailable for certain times pose no particular problems for a Bayesian analysis. At any time for which there is no observation on the series of interest, one's posterior distribution on model parameters is straightforwardly one's prior for that time. With no new information becoming available, estimates and uncertainties remain unaltered. This is the basic situation. Irregularly observed time series can be handled in the same way by defining an equally spaced underlying time scale, placing the known observations appropriately on that scale, and treating the unfilled times as missing values. BATS is quite capable of dealing with missing observations: this chapter explains and demonstrates the mechanics.

12.1 Communicating Missing Values

Missing values are represented in data files by the string **N/A** which stands for 'Not Available'. The same representation is used when displaying time series—either data or model estimates—using the **Table** facility. One may also set missing values using the data editor, again using **N/A** for communication. Let's take a look at these operations.

We will load the data first seen in Chapter 10, stored in the **CANDY.DAT** data file. If you are beginning this tutorial 'from scratch', you will first need to start the program. The necessary startup instructions are detailed at the beginning of Chapter 9 if you need reminding. We shall assume that you have done this and that the root menu is currently displayed. Go ahead and load the data.

▷ From the root select in turn **Data**, **Input-Output**, **Read**, **BATS-format**.

The screen should now be displaying the **.DAT** files in your default directory (unless you have defined your own files and directories configuration as described in the Appendix to Chapter 9). If **CANDY.DAT** is not included in the list then you will have to press **Page Down** and specify a full pathname for the file. We shall assume the default configuration in the instructions here.

▷ Using the arrow keys highlight **CANDY.DAT** and press **ENTER**.
▷ Quit from the **Read** menu.

Now we will use the data editor to set some missing values.

▷ Select **Edit** (from the **Data/Input-Output** menu).

The table display that we have seen before now appears. Notice the **[Edit]** label at the top of the column of row labels, and the blinking block cursor in the first (highlighted) cell. These indicate that the table is in edit mode and so the contents may be changed. Contrast this with our usage of tables for displaying the results of analyses in previous tutorials. There, the table was in 'view' mode: although we could move around the table, no cursor was present and no changes could be made to the displayed values. For the illustration here we will change the **SALES** entries for 1973/3 and 1979/12 to missing values.

▷ Press the **right arrow** key to move the cursor (and highlight) to the first entry in the **SALES** column.
▷ Press the **down arrow** key repeatedly until the cursor is positioned at the 1977/3 value for **SALES** (10.57).

12.1 Communicating Missing Values

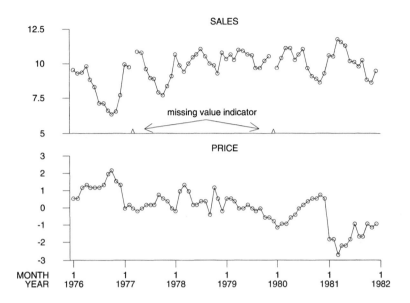

Figure 12.1 Time plot of CANDY series SALES (modified) and PRICE.

▷ Type **N**, then press the **ENTER** key.

See that the **SALES** value for 1977/3, 10.57, has been replaced by the missing value indicator, **N/A**. Now repeat the procedure to set a missing value for **SALES** in 1979/12.

  ▷ Press the **Page Down** key twice to bring 1979/12 within the viewing area. The **SALES** entry for 1980/7 (10.54) is currently highlighted.
  ▷ Press the **up arrow** key repeatedly to move the cursor to 1979/12.
  ▷ Type **N**, then press the **Enter** key.
  ▷ Press **ESC** to complete the editing procedure, returning control to the **Data/Input-Output** menu.

The values for **SALES** in March of 1973 and December of 1979 have now been deleted and made known to BATS as missing. Let's take a look at what this special status implies for a graphical display.

  ▷ Quit to the **Data** menu.
  ▷ Select **Explore**, then **MPlot**.
  ▷ Press **ENTER** to mark **PRICE** for selection.
  ▷ Press the **down arrow** key to highlight **SALES**, press **ENTER** then **ESC**.

A time plot of the **PRICE** and newly modified **SALES** series is displayed (Figure 12.1). You can see from the upper frame (which displays the **SALES**

Table 12.1 CANDY series SALES (modified), and PRICE

Time	PRICE	SALES
1976/1	0.536	9.535
1976/2	0.536	9.245
1976/3	1.122	9.355
1976/4	1.321	9.805
1976/5	1.122	8.821
1976/6	1.122	8.321
1976/7	1.122	7.118
1976/8	1.321	7.100
1976/9	1.927	6.593
1976/10	2.133	6.311
1976/11	1.521	6.564
1976/12	1.321	7.732
1977/1	−0.035	9.929
1977/2	0.154	9.750
1977/3	−0.035	N/A
1977/4	−0.222	10.840

series) that no points are drawn for those times that we have just set to missing, with the result that the line joining the observations is split into three disjoint segments. Notice also that a small marker is drawn just above the time axis to indicate each of the missing values. These two features are present whenever a graph displays a series that contains missing values. Graphs of other series are not affected as you can see from the lower frame in which the PRICE series is displayed. Let's now take a look at a table of the series values.

▷ Press Q to quit the time plot.
▷ Select Table.
▷ Highlight PRICE, then SALES and press ENTER for each; press ESC.

The table display appears—in view mode this time—and you can see that SALES for 1977/3 and 1979/12 are marked as missing (Table 12.1).

▷ Press ESC to quit the table.
▷ Quit to the root.

12.2 Summary of Operations

This summary covers the basic notions of working with missing values; data entry and graphical indicators.

1. Loading the example data file
   - From the root select in turn Data, Input-Output, Read, BATS-format.
   - Using the arrow keys highlight CANDY.DAT and press ENTER.
   - Quit from the Read menu.
2. Editing data: setting missing values
   - Select Edit (from the Data/Input-Output menu).
   - Press the right arrow key to move the cursor (and highlight) to the first entry in the SALES column.
   - Press the down arrow key repeatedly until the cursor is positioned at the 1977/3 value for SALES (10.57).
   - Type N, then press the ENTER key.
   - Press the Page Down key twice to bring 1979/12 within the viewing area. The SALES entry for 1980/7 (10.54) is currently highlighted.
   - Press the up arrow key repeatedly to move the cursor to 1979/12.
   - Type N, then press the Enter key.
   - Press ESC to complete the editing procedure, returning control to the Data/Input-Output menu.
3. Viewing modified series
   - Quit to the Data menu.
   - Select Explore, then MPlot.
   - Press ENTER to mark PRICE for selection.
   - Press the down arrow key to highlight SALES, press ENTER, then ESC.
   - Press Q to quit the time plot.
   - Select Table.
   - Highlight both PRICE and SALES and press ENTER for each; press ESC.
   - Press ESC to quit the table.
   - Quit to the root.

12.3 Analysis with Missing Values I: Response Series

Missing values in the response series are easy to deal with in Bayesian forecasting and time series analysis. Forecasting, it should be clear, is not affected by the response series—these are the values being predicted! Monitoring forecast performance and updating estimates are affected of course. With no observed value there is no forecast error defined, so there is nothing by which to judge forecast performance. Monitoring is effectively suspended for such times. Learning is also suspended because there is no incoming information from which to learn. In the absence of any subjectively included information, posterior distributions on model parameters are just the existing prior distributions and the analysis rolls forward to the next time. (This is precisely the state of affairs we saw in Chapter 11 when a response was treated as an outlier following a monitor signal. So you already know how it works!) To see this in practice let us model the SALES series using a steady trend and a regression on PRICE. (You may recall that we analysed this model in Chapter 10 and you may like to compare the relevant results there to those we are about to see here.

 ▷ From the root select Model.
 ▷ Set SALES as the response: press the down arrow key to highlight SALES then press ENTER.
 ▷ Select Trend, then Constant to define a steady trend component.
 ▷ Select Regression; with PRICE highlighted press ENTER, then ESC to define a regression on PRICE.
 ▷ Quit to the root.

Since the focus of our interest here is on missing value handling we will just go ahead with an analysis using the program default discount factors and reference initialisation. You may, of course, set whatever component discounts and prior moments that you wish, but such changes do not affect the missing value procedures.

 ▷ Select Fit, then Reference.

The forecast graph (Figure 12.2) illustrates some of the special features for missing values that we described for the time plot above. Unobserved SALES values at 1977/3 and 1979/12 are indicated by markers drawn on the time axis, but notice that the line joining the point forecasts is *not* broken. Since we do not need an observation on the response series in order to make a forecast, there are no missing forecasts and hence no broken line. There is one modification to the display that we have not seen before: a list of the missing value times displayed in the model summary panel. Information is only displayed in the summary panel when a model is defined. When we looked at the time plot of SALES and PRICE above we did so before defining a model, and so the missing values were not listed there. (You may recall

12.3 Analysis with Missing Values I: Response Series

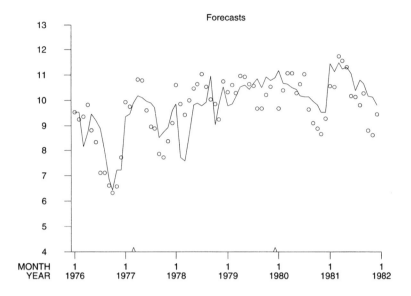

Figure 12.2 One step forecasts with missing responses.

from Chapter 11 that outliers and forward interventions are also listed in this way. In practice, you may have combinations of all three kinds of 'special' times listed in the summary panel.) If you now perform the back filtering analysis, you will see similar features in the retrospective graph.

▷ Press **Q** to quit the forecast graph.
▷ Select **Retrospective**.

There is really nothing more to say about the retrospective analysis with regard to missing values than has already been said for the on-line forecasts, so let's pass on immediately to take a look at some analysis results. We begin by examining the fitted residuals series.

▷ Press **Q** to quit the retrospective graph.
▷ Select **Data**, **Explore**, then **Time-Plot**.
▷ Highlight **S Std Residuals**, press **Enter**, then **ESC**.

When a response series observation is missing it is not possible to define a residual—forecast, fitted, or otherwise—and so the residuals graph (Figure 12.3) takes on the broken appearance that we saw in Figure 12.1 for the **SALES** time plot. There are no missing value indicators drawn on the time axis in this case because no points are plotted.

You may be wondering at the high frequency of very large standardised fitted residuals displayed here. Clearly the model as specified is rather

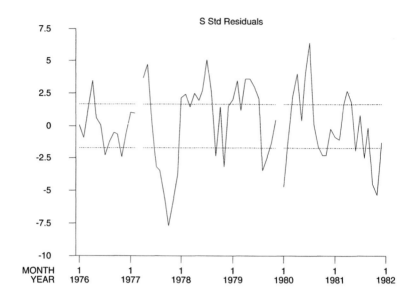

Figure 12.3 Time plot of fitted residuals with missing responses.

inadequate! There is a good explanation for this of course: our aim in this chapter is to demonstrate missing value handling not model building, and the model we are using has been chosen purely for convenience. A more appropriate model for this data was built (in a different but still illustrative context) in the latter part of Chapter 10.

Table 12.2 provides another view of the fitted residuals series. As you can see, the residuals for 1977/3 and 1979/12 are represented—as expected since they are undefined—by the missing value indicator, N/A.

 ▷ Press Q to quit the residuals time plot.
 ▷ Select Table.
 ▷ Highlight SALES, Fitted, and S Std Residuals and press ENTER for each in turn.
 ▷ Press ESC to see the table.

When you have finished examining the table, return up the menu hierarchy to the root.

 ▷ Press ESC to quit the table.
 ▷ Quit to the root.

Table 12.2 Sales, fitted values, and fitted residuals with missing responses

Sales	Fitted Values	90%	S Std Res	90%
9.535	9.518	0.572	0.049	1.675
9.245	9.509	0.484	−0.915	1.675
9.355	8.972	0.493	1.300	1.674
9.805	8.736	0.516	3.467	1.674
8.821	8.647	0.491	0.589	1.674
8.321	8.305	0.531	0.049	1.674
7.118	7.942	0.606	−2.277	1.673
7.100	7.547	0.603	−1.239	1.673
6.593	6.792	0.624	−0.534	1.673
6.311	6.563	0.630	−0.670	1.673
6.564	7.357	0.550	−2.411	1.673
7.732	7.891	0.520	−0.512	1.672
9.929	9.663	0.444	0.998	1.672
9.750	9.532	0.385	0.944	1.672
N/A	9.769	0.386	N/A	N/A
10.840	9.998	0.384	3.662	1.672

12.4 Analysis with Missing Values II: Regressor Series

As we have just seen, the treatment of missing values in the response series is quite straightforward. There is no special action which must be taken either with respect to model definition or to analysis. One simply needs to recognise the representations of missing values—the N/A string in tables, and the graphical markers, broken lines, and time lists. Missing regressors are dealt with in a similar manner but there is an additional analytical complication. When a regression variable is not available at some time it is not possible to define model forecasts or fitted values for that time. This contrasts with what we saw for the response series. A missing response evoked the modelling equivalent of a shrug of the shoulders, or a 'never mind' attitude; the reaction to missing regressors is more like a scratching of the head, or 'hmmm....we shall just have to wait and see'. Since the model cannot produce a forecast there is no way to define a residual when the response is observed and therefore it is not possible to apportion information in the response among the model components. In other words, as with the case of unobserved response there is no updating of parametric distributions at times of missing regressors. The priors become directly the posteriors (in the absence of other subjective information). It is, of course, entirely feasible to make a subjective assessment of the implications of the observed response. For example, it may be possible to assume a certain

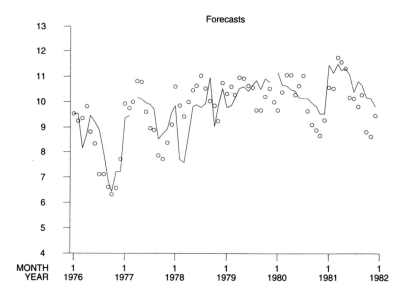

Figure 12.4 One step forecasts with missing regressors.

interval in which the unknown regressor should lie and thereby deduce bounds on the partitioning of information from the response among the model components. Whatever subjective assessments are made, the resulting posterior parameter estimates and uncertainties may be communicated to the program as an intervention as described in Chapter 11.

Let us see the treatment of missing regressors in practice. We will remove the missing SALES values and add some missing PRICEs.

▷ From the root select **Data**, then **Input-Output**, then **Edit**.
▷ Using the arrow keys move the cursor to the 1977/3 cell for PRICE.
▷ Type **N**, then press **ENTER**.
▷ Press the **right arrow** key to move the cursor to the corresponding SALES cell.
▷ Type 10, then press **Enter**.
▷ Using **Page Down** and the **up/down/left/right** arrow keys repeat the above procedure for SALES and PRICE in 1979/12.
▷ Press **ESC** to finish the data editing.
▷ Quit to the root.
▷ Select **Fit**, then **Reference**.

This time the forecast graph is broken at the missing regressor times (Figure 12.4). As we have said, no regressor value means no forecast, hence the breaks. Notice that there are no missing value indicators drawn on the

12.4 Analysis with Missing Values II: Regressor Series

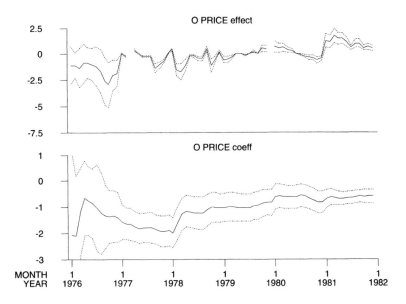

Figure 12.5 On-line fitted regression estimates with missing regressors.

time axis. This is because it is regressors that are missing—easily apparent from the breaks in the line joining the point forecasts—the response series is complete. In the case where the response is missing in addition to a regressor, an indicator *is* drawn. The times of the breaks are listed in the summary panel as before (not shown in Figure 12.4). These same features will be evident in the retrospective graph, which we shall not bother to examine this time around. Instead, let us take a look at the on-line regression estimates to see what effects missing regressors induce there. We will inspect both the **PRICE** variable regression coefficient and the effect of the regression on the analysis.

▷ Press **Q** to quit the on-line forecasts graph.
▷ Select **Data**, **Explore**, **MPlot**.
▷ Highlight **O PRICE Coeff** and **O PRICE Effect** and press **ENTER** for each in turn; press **ESC**.

As you can see from Figure 12.5 the regression coefficient (in the lower frame of the graph) has a properly defined posterior distribution, represented on the graph by the posterior mean and 90% uncertainty limits, for all times—including those times for which the regressor itself is missing. As we have already said, posterior distributions for model parameters are well defined regardless of the present/absent status of regressor variables (or the response). The effect of the regressor, in contrast, is not defined if the regressor is missing since it is the product of the coefficient and the regression value—which is not available when it is missing! You can see this from the breaks in the estimated PRICE effect shown in the upper frame of the graph.

When you have finished with the regression estimates graph return to the root.

▷ Press Q to quit the regression estimates graph.
▷ Quit to the root.

12.5 Summary of Operations

This summary covers the model definition and illustrative missing data analyses. The SALES series is modelled with a steady trend and a regression on PRICE.

1. Defining the model
   - From the root, select **Model**.
   - Set SALES as the response: press the **down arrow** key to highlight SALES, then press ENTER.
   - Select **Trend**, then **Constant** to define a steady trend component.
   - Select **Regression**; with PRICE highlighted press ENTER, then ESC to define a regression on PRICE.
   - Quit to the root.
2. Missing response analysis
   - Select **Fit**, then **Reference**.
   - Press Q to quit the forecast graph.
   - Select **Retrospective**.
   - Press Q to quit the retrospective graph.
   - Select **Data**, **Explore**, then **Time-Plot**.
   - Highlight S Std Residuals, press Enter, then ESC.
   - Press Q to quit the residuals time plot.
   - Select **Table**.
   - Highlight SALES, Fitted, and O Std Residuals and press ENTER for each in turn; press ESC to see the table.
   - Press ESC to quit the table.
   - Quit to the root.
3. Missing regressor analysis
   - From the root, select **Data**, then **Input-Output**, then **Edit**.
   - Using the arrow keys move the cursor to the 1977/3 cell for PRICE.
   - Type N, then press ENTER.
   - Press the **right arrow** key to move the cursor to the corresponding SALES cell.
   - Type 10, then press Enter.
   - Using **Page Down** and the **up/down/left/right** arrow keys repeat the above procedure for SALES and PRICE in 1979/12.
   - Press ESC to finish the data editing.
   - Quit to the root.
   - Select **Fit**, then **Reference**.
   - Press Q to quit the on-line forecasts graph.
   - Select **Data**, **Explore**, **MPlot**.
   - Highlight O PRICE Coeff and O PRICE Effect and press ENTER for each in turn; press ESC.
   - Press Q to quit the regression estimates graph.
   - Quit to the root.

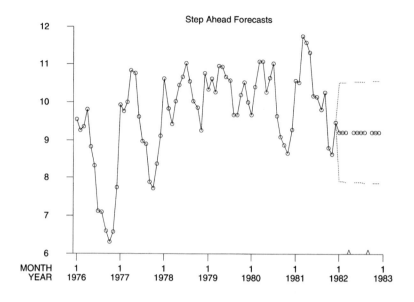

Figure 12.6 Prediction with missing regressors.

12.6 Prediction with Missing Regressors

Calculating k step ahead predictions is done in much the same way as the one step forecasts already discussed. That is to say, for those times at which model regressors are missing no forecast is defined. To see this in BATS we will calculate forecasts for a couple of years following the analysis just completed.

▷ From the root select **Predict**.
▷ Enter **1982/12** for the forecast horizon; press **ESC**.

As we saw in Chapter 10, when predicting beyond the end of the data with models that include regression components we have to enter values for the regressors over the 'extension' period. This is the interval from the end of the data to the specified forecast horizon, January 1981 to December 1982 in our example. Regressor specification is done using the table display which explains why the table has appeared at this time. BATS initialises the regressors over the extension period with zero; for our illustration we want to include some missing values.

▷ Press **Page Down** four times to go to the end of the extended data set.
▷ Using the arrow keys highlight the cell for PRICE 1982/4, then type **N**; press **ENTER**.
▷ Repeat this procedure to set a missing value for PRICE at 1982/9.

12.8 Data Transformations with Missing Values

▷ Press ESC to complete the editing.
▷ Select Continue.

The prediction graph appears (Figure 12.6). Notice that it combines all of the missing data annotations we have described: 'broken' forecasts, time index markers, and a summary list (which is not shown in the figure). Exit the graph when you are ready to continue with the tutorial.

▷ Press Q to quit the prediction graph.
▷ Quit to the root.

12.7 Summary of Operations

This summary covers forecasting beyond the data end time following an on-line analysis.

1. Set the forecast horizon
   ▷ From the root, select Predict.
   ▷ Enter 1982/12 for the forecast horizon; press ESC.
2. Set future missing PRICEs
   ▷ Press Page Down four times to go to the end of the extended data set.
   ▷ Using the arrow keys highlight the cell for PRICE 1982/4, then type N; press ENTER.
   ▷ Repeat this procedure to set a missing value for PRICE at 1982/9.
   ▷ Press ESC to complete the editing.
3. Examine the predictions
   ▷ Select Continue.
   ▷ Press Q to quit the prediction graph.
   ▷ Quit to the root.

12.8 Data Transformations with Missing Values

Missing values are transformed to missing values—it is as simple as that. Let us see by example. The PRICE series currently contains missing values for 1977/3 and 1979/12. (If you have been following the instructions in this tutorial so far; if not, simply set some missing values before proceeding—the first summary of commands in this chapter will help you if you require assistance with this.) We will create a series of logarithmically transformed prices. (Chapter 10 introduced the data transformation/series creation facility; we assume that you understand what is going on and will not describe it again here.)

▷ From the root select Data, Input-Output, Create.

Table 12.3 Transformations with missing values

PRICE	logprice
0.536	−0.622
0.536	−0.622
1.122	0.115
1.321	0.278
1.122	0.115
1.122	0.115
1.122	0.115
1.321	0.278
1.927	0.655
2.133	0.757
1.521	0.419
1.321	0.278
−0.034	N/A
0.154	−1.870
N/A	N/A
−0.222	N/A

▷ Enter **logprice** for the target series name.
▷ Press the **down arrow** key to move to the formula box; enter **loge Price** to apply the natural log function to price.
▷ Press **ESC** to complete the transformation specification.

You can see from the data description now displayed that the **logprice** series has been created. Take a look at it to confirm the handling of missing values under transformation.

▷ Quit to the **Data** menu (two steps up the hierarchy; if you get lost, recall that **Data** is an option on the root menu).
▷ Select **Explore**, then **Table**.
▷ Highlight **PRICE**, then **logprice** and press **ENTER** for each in turn; press **ESC**.

If you inspect the entries for 1977/3 you will see that the missing value for **PRICE** has indeed been transformed to a missing value for **logprice** (Table 12.3). But more than this you can see that nonpositive **PRICE**s have also been transformed to missing **logprice**s. This is an example of a general rule in data transformations: whenever a transformation function is applied to an argument outside of that function's domain (such as nonpositive values for logarithms in the current illustration) the result is defined to be a missing value. This is done automatically and no prompt or warning is given. Beware!

12.10 Items Covered in This Tutorial   **357**

That brings us to the end of what we have to say about missing data. To finish the tutorial, exit from the table, then from the program.
- ▷ Press ESC to quit the table.
- ▷ Quit to the root.
- ▷ Select Exit, then Yes.

12.9 Summary of Operations

This section showed how data transformations deal with missing values and also how these transformations deal with values outside their defined domain. The demonstration assumes that the PRICE variable for 1977/3 is set to missing.

1. Creating the transformed series
   - ▷ From the root select Data, Input-Output, Create.
   - ▷ Enter logprice for the target series name.
   - ▷ Press the down arrow key to move to the formula box; type loge Price to apply the natural log function to price.
   - ▷ Press ESC.
2. Examining the transformed series
   - ▷ Quit to the Data menu (if you get lost, recall that Data is an option on the root menu).
   - ▷ Select Explore, then Table.
   - ▷ Highlight PRICE and logprice and press ENTER for each in turn; press ESC.
3. Finishing up
   - ▷ Press ESC to quit the table.
   - ▷ Quit to the root.
   - ▷ Select Exit, then Yes to quit the session.

12.10 Items Covered in This Tutorial

In this tutorial we have seen how missing values in time series are handled in Bayesian analysis: implications for prediction and estimation; contrasting effects of missing values in the response series and in regressor series.
- Communicating missing values: N/A in data editing and display.
- Graphical indicators: markers, broken lines, and summary list.
- Forecasting with missing values.
- Analysis with missing values in response and/or regressor.
- Model estimates from missing value analysis.
- Transformations of missing values.

Chapter 13

Tutorial: Data Management

This tutorial describes how to set up your files in a form in which BATS can understand the contents, how to manipulate data series by selecting subsets for analysis, how to append additional values to series, how to create aggregates from 'low level' series, and how to save and restore model definitions with data sets. Transformations of individual time series and direct data editing have already been discussed in Chapter 10.

There are two alternative file formats that can be used to communicate data to the program. Both are plain text formats, so all BATS data files may be created, inspected, or modified using standard text editors. The simplest of the two formats is the so-called 'free format' which defines a file structure for just raw numbers. The other is the 'BATS format' which defines a file structure for time series data and data set descriptive information. It is most likely that you will initially communicate your data to BATS using the free format file structure—perhaps created with some other application program that you use. For subsequent usage it is more convenient to use the BATS format since this obviates the step of separately specifying the necessary timing information. The model save/restore facility provides no option: the BATS format is always used.

13.1 Free Format Data Files

The free format data file has a particularly simple and quite standard form. Table 13.1 shows a short example. This example contains two series

Table 13.1 Free format data file

0.536	9.535
0.536	9.245
1.122	9.355
1.321	9.805

of length four, one series per column (with columns separated by white space—spaces or tab characters), one set of observations per row. This is the usual format in which new data sets will be first communicated to BATS. The file contains only the raw time series values and nothing more—no titles, no labels, no descriptive information of any kind.

13.1.1 Reading the File

Raw lists of numbers are not sufficient for modelling with BATS; the program requires information about the timing characteristics of the data: specifically, a starting time, a cycle length or period, and labels for the timing cycle and timing interval. Names for the individual series are not actually necessary—but they will certainly make your life easier! This information has to be supplied separately when a free format data file is loaded.

The data in Table 13.1 are successive quarterly recordings of sales and price of a particular commodity over the period from quarter three, 1989 to quarter two, 1990. The data is contained in the example data file FREE1.FRE. Load the file and tell BATS the timing and naming information.

▷ From the root select Data, Input-Output, Read, Free-format.
▷ Using the arrow keys highlight FREE1.FRE.
▷ Press ENTER.

(If the example data file FREE1.FRE does not appear in the list of files displayed, then either the file itself is not contained in the directory from which BATS was started, or you have configured a free format data file directory/mask combination which does not contain the file. In either case you can press the Page Down key while the file list is displayed and enter a full path name for FREE1.FRE in the resulting dialogue box. Directory configurations are described in the appendix to Chapter 9, and the pathname file specification procedure is described in Chapter 9 itself.)

A screen form appears with dialogue boxes for entering the details of the data set. We will take each box in turn. The title and source boxes are for purely descriptive information. The information you enter here will only be used in the data set descriptions displayed during the session. The cycle name and timing name are used in graphical displays in addition

13.1 Free Format Data Files

to data set descriptions. The **period** and **starting cycle/time** are the basic timing information fundamental to the description of a time series. This information is used in data set descriptions, graphs, and modelling. Finally, the **series names** are used for individual series identification and selection, for model components, graphs, and tables. Fill out the form as follows.

▷ Enter **example** for the title.
▷ Press the **down arrow** key and enter **user guide** for the source.
▷ Press the **down arrow** key and enter **year** for the cycle name.
▷ Press the **down arrow** key and enter **quarter** for the timing name.
▷ Press the **down arrow** key and enter **4** for the period.
▷ Press the **down arrow** key and enter **1989** for the start cycle.
▷ Press the **down arrow** key and enter **3** for the start time.
▷ Press the **down arrow** key and enter **sales** for the first series name.
▷ Press the **down arrow** key and enter **price** for the second series name.
▷ Press **ESC** to have BATS read the form.

(The backspace, space, and arrow keys are useful if you make a mistake while typing and need to correct it.) The data set description shows all of the information just entered, and also some additional information. The number of observations was determined by the program from the file contents, and this is used to calculate the end cycle/time given the start cycle/time and period that we entered.

▷ Quit to the root.

13.1.2 Writing the File

We said in the introduction to this tutorial that, while the free format is particularly useful for initially communicating your data to BATS, subsequent use is more convenient if the BATS format is used. Imagine having to go through the steps we have just taken with **FREE1.FRE** every time we wanted to use this data. Well, of course there is no need to do this. Once series names and timing details are known they may be stored on file along with the series values by using the BATS file format. Typically, having loaded a new data set in the free format as we have just done for **FREE1.FRE**, one would immediately write it out again using the BATS format. Then, whenever this data set is required, the series values and descriptive information can be read from file: no other input is required from us. This is how the data sets used in the previous tutorials were loaded if you recall.

If you follow the procedure of always saving data sets in the BATS format, you may be wondering if it is necessary to keep the free format copy around.

The answer to that is no. If you ever do want the data unencumbered with descriptive information, then just load it into BATS and write it out again using the free format.

One more note: the free format file structure is also used when writing the contents of tables to file.

13.2 BATS Format Data Files

The BATS format for data files is basically an extension of the free format just discussed. At the beginning of the file—before the series values—the details of series names and timing information are recorded. A 'delimited' line structure is used, which simply means that every line in the file begins with a 'delimiting' character—in fact a backslash, \. We won't bother you with the details of why this is done; it is sufficient for present purposes just to note that it is done.

Following the descriptive information come the series values. These are arranged in exactly the same way as in the free format files: one series per column, columns white space separated, and one set of observations (one time point) per row.

Additional comments may be included following the last row of data; such comments will be ignored by the program. (So if you later overwrite a BATS format file, any comments in the original file will be lost.) Building BATS format data files is not something one would normally do 'by hand', so we will not discuss the format details further at this time.

13.2.1 Storing and Retrieving Model Definitions

Model definitions may be stored on file along with the relevant data set. BATS does this by appending the model details to a BATS format data file. The details recorded are the designated response series, the model component definitions, and the individual component discount factors for each time.

We will go through the steps involved in writing a model definition to file and then restoring that definition to see how things work. With our example data set still present, define a constant trend model for the price variable.

▷ From the root select **Model**.
▷ Press **ENTER** (with the default first entry, **price**, highlighted) to set **price** as the response.
▷ Select **Components**.
▷ Select **Trend**, then **Constant**.

13.2 BATS Format Data Files

▷ Quit to the root.

Now write the data and model definition to file.

▷ Select Data, Input-Output, Write.
▷ Select Write-Model.
▷ Press Page Down and enter example.dat in the dialogue box (we do not want to overwrite any of the other data files!).
▷ Press ESC to complete the write operation.

To demonstrate that the saved model definition is properly restored from file we will first discard the current model and data set.

▷ Quit (twice) to the Data menu.
▷ Select Clear.

The screen display shows that the data set has been erased. A consequence of the erasure is a contraction of the Data menu: the Explore and Clear options are no longer available because with no data present there is nothing to explore and nothing to clear. This contraction shows the conditional option mechanism working 'in reverse' compared with what we have seen before. In previous tutorials we have only seen menus grow; we have now seen that menus can shrink too.

Loading a model definition is similar to loading a data set. On the Read menu one selects the Read-Model option instead of the BATS-format option we have used before.

▷ From the (currently active) Data menu, select Input-Output, Read.
▷ Select Read-Model.

We get the list of BATS format data files (files with extension .DAT by default)—we want EXAMPLE.DAT.

▷ Use the arrow keys to highlight EXAMPLE.DAT, then press ENTER.

BATS tells us that it has located a model definition on the data file and wants to know if we would like to load it. If we say 'No', the model details are ignored leaving just the data set loaded. We do want to load the model.

▷ Select Yes.

Voila! The constant trend model for price in the example data set has been restored.

▷ Quit to the root.

13.3 Summary of Operations

This summary covers the loading of free format data files and subsequent entering of descriptive information, saving model definitions to file, and restoring model definitions from file.

1. Loading a free format data file
   - From the root select **Data, Input-Output, Read, Free-format**.
   - Using the arrow keys highlight **FREE1.FRE**.
   - Press **ENTER**.

2. Specifying timing and naming information
   - Enter **example** for the title.
   - Press the **down arrow** key and enter **user guide** for the source.
   - Press the **down arrow** key and enter **year** for the cycle name.
   - Press the **down arrow** key and enter **quarter** for the timing name.
   - Press the **down arrow** key and enter **4** for the period.
   - Press the **down arrow** key and enter **1989** for the start cycle.
   - Press the **down arrow** key and enter **3** for the start time.
   - Press the **down arrow** key and enter **sales** for the first series name.
   - Press the **down arrow** key and enter **price** for the second series name.
   - Press **ESC** to have BATS read the form.
   - Quit to the root.

3. Saving a model definition to file (i) define a model
   - From the root select **Model**.
   - Press **ENTER** (with the default first entry, **price**, highlighted) to set **price** as the response.
   - Select **Components**.
   - Select **Trend** then **Constant**.
   - Quit to the root.

3. Saving a model definition to file (ii) write the model to file
   - Select **Data, Input-Output, Write**.
   - Select **Write-Model**.
   - Press **Page Down** and enter **example.dat** in the dialogue box (we do not want to overwrite any of the other data files!).
   - Press **ESC** to complete the write operation.

4. Clearing an existing data set from the program
   - Quit (twice) to the **Data** menu.
   - Select **Clear**.

5. Loading a data set and model definition from file
   - From the (currently active) **Data** menu, select **Input-Output, Read**.
   - Select **Read-Model**.
   - Use the **arrow** keys to highlight **EXAMPLE.DAT**, then press **ENTER**.
   - Select **Yes** (to retrieve the model details).
   - Quit to the root.

13.4 Subset Selection

If you wish to analyse only a portion of your data set, BATS will allow you to discard values from the beginning and from the end by redefining the data set start cycle/time and end cycle/time. Load the CANDY.DAT data set so we can illustrate the selection procedure.

▷ From the root select Data, Input-Output, Read, BATS-format.
▷ Using the arrow keys highlight CANDY.DAT.
▷ Press ENTER.

(Our usual caveat about file and directory configurations applies here. The foregoing load instructions assume that the CANDY.DAT data file exists in the configured data directory.) The selection utility is part of the Data/Input-Output menu which is one level up from the current Read menu.

▷ Quit from the Read menu to the Data/Input-Output menu.
▷ Choose Select.

As you can see, the current data set timing information is displayed in the control portion of the screen and dialogue boxes are presented for respecifying the starting and ending cycles and times. To illustrate, let us discard the first and last year of the data.

▷ Enter 1977 for the new start cycle.

The cursor automatically jumps to the next box in the group when the end of the current box is reached. We are not going to alter the 'start time' value (the 'one' corresponds to January since this is monthly data) so move immediately to the 'new end cycle' box.

▷ Press the down arrow key to move to the 'new end cycle' box.
▷ Enter 1990 for the 'new end cycle'.
▷ Press ESC to apply the selection.

Notice what the request for a new end year of 1990 means. It is beyond the time of the last recorded observation of the present data set. But this is not allowed because no data values are defined there. Hence we are politely asked if we wish to abort the selection altogether (retaining the data set intact) or to try again. Let us try again.

▷ Select No (we do not wish to abort the selection procedure).

The dialogue boxes return, filled with the timing entries we have just requested—including the invalid end time. We will correct the end time—setting it to one year before the existing end time as we said above.

▷ Press the down arrow key twice to move to the 'new end cycle' box.
▷ Enter 1980 for the 'new end cycle'.
▷ Press ESC to apply the selection.

Success this time! The data set description displays the new timing information and tells us that there are now only 48 observations. (There were 72 to begin with and we have lopped off one year's worth—12—from both the beginning and the end.) If you experiment with displaying the data—in graphs or tables from the **Explore** menu—you will see that indeed the original first and last year's data are no longer present.

A word of warning: subset data selection does *not* modify data files. If you wish to store the (shortened) data set, you must explicitly write it to file.

13.5 Temporal Aggregation

Data is not always readily available on a timing scale suitable for a particular analysis. And one may often want to analyse data using different time scale granularities. For example, data may be available on a monthly basis and one may want to build a quarterly model. BATS provides a mechanism for aggregating data series in this way. On the **Data/Input-Output** menu currently displayed you can see the **Aggregate** option. We shall investigate this by converting the monthly **CANDY.DAT** data set (currently reduced to the four years from 1977 to 1980 inclusive following the previous exercise) into a quarterly series.

▷ Select **Aggregate**.

A set of eight dialogue boxes appears; the entries made here determine exactly how the time series will be aggregated. The first two boxes allow you to set the starting position for the aggregation—the default is always the current starting point of the data set. On this occasion the default is fine, but consider the situation if our monthly records began in February rather than January. With conventionally defined quarterly intervals this would mean that we should begin the aggregation in April and ignore the values for February and March. BATS will not object if you ask to begin aggregation in February (or any other month) so you must be careful. A similar warning goes for any other timing scale that you may be aggregating over.

The content of the third box, **Number of points to cumulate**, determines the extent of the aggregation. There are three months in a quarter so the appropriate value to enter here is 3. Once again, BATS will not object to any *feasible* value that you enter. If you want to cumulate your monthly data in batches of 13 months and call it 'quarterly', then as far as the program is concerned you may do so. Objections will be raised, of course, if you should ask to cumulate 0 months or more months than there are existing observations, or specify a timing period less than 1.

▷ Press the **down arrow** key twice to move to the **Number of points to**

13.5 Temporal Aggregation

cumulate box.
▷ Enter 3.

There are four quarters in a year so the next entry (data period) is obvious.

▷ Press the down arrow key twice to move to the New data period box.
▷ Enter 4.

Quarter one for 1977 is the new starting point.

▷ Press the down arrow key twice to move to the New start cycle box.
▷ Enter 1977, then 1 (the cursor jumps automatically to the time box when you reach the end of the cycle box).

Quarterly data is not usefully labelled as 'month/year' and in general either or both timing labels may be inappropriate for the aggregated scale. So the last two dialogue boxes allow these labels to be reset. Once again our entries are obvious.

▷ Press the down arrow twice to move to the New cycle names box.
▷ Enter year for the 'new' timing cycle.
▷ Press the down arrow key and enter quarter for the new timing interval.
▷ Press ESC to apply the changes.

The data set description shows the result of the data aggregation. The original 48 observations are now reduced to just 16 beginning in quarter one, 1977, and ending in quarter four, 1980.

13.5.1 Aggregation and Structural Change

An important point to be aware of when aggregating data series is the effect of a structural change in the series. Any change at a timing interval on the original scale that falls part way through an interval on the aggregate scale will result in a 'halfway house' aggregate value. On the aggregate scale the underlying change will appear as a transition taking two intervals. For example, suppose that sales are running at 100 units from January through May and subsequently jump to 150 units in June. Now we aggregate the series to the quarterly scale yielding sales figures of 300, 350, 450, 450. As you can see, without investigating the more detailed series we could easily be lead to draw the conclusion of a six monthly upward drift in sales, rather than the more accurate sharp jump in June. Incidentally, this point should be borne in mind when examining any data set: it may already be recorded on an aggregated scale before you get to it.

There are various ways of dealing with the problem. If the starting point for data aggregation is not important, shift it to ensure that the (probable) time of change occurs at the beginning of an aggregate interval. If the timing interval definition is important—as it very likely will be in the kind of commercial example above—then one must recognise the problem exists and make appropriate adjustments to the analysis performed and conclusions drawn. In essence we are really advising you to be careful.

13.6 Editing and Transformations

BATS provides a built in data editor and a simple series transformation utility. The data editor is used in several places in the program. We have seen it used in earlier tutorials to change raw data values, to set outlier and intervention indicators, to set future regressor values for predictions, and (in view-only mode) to examine the results of analyses. We have also seen the operation of the series transformation utility: in Chapter 10 we created the logarithmic transformation of the PRICE variable in the CANDY.DAT data set.

13.7 Summary of Operations

This summary covers data set subset selection, data set extension, and data aggregation.

1. Subset selection
   - From the root select Data, Input-Output, Read, BATS-format.
   - Using the arrow keys highlight CANDY.DAT.
   - Press ENTER.
   - Quit from the Read menu to the Data/Input-Output menu.
   - Choose Select.
   - Enter 1977 for the new start cycle.
   - Press the down arrow key to move to the new end cycle box.
   - Enter 1990 for the new end cycle.
   - Press ESC to apply the selection.
   - Select No (we do not wish to abort the selection procedure).
   - Press the down arrow key twice to move to the new end cycle box.
   - Enter 1980 for the new end cycle.
   - Press ESC to apply the selection.
2. Data aggregation
   - From the Data/Input-Output menu, select Aggregate.
   - Press the down arrow key twice to move to the Number of points to cumulate box.
   - Enter 3.
   - Press the down arrow key twice to move to the New data period box.
   - Enter 4.
   - Press the down arrow key twice to move to the New start cycle box.
   - Enter 1977, then 1 (the cursor jumps automatically to the time box when you reach the end of the cycle box).
   - Press the down arrow twice to move to the New cycle names box.
   - Enter year for the 'new' timing cycle.
   - Press the down arrow key and enter quarter for the new timing interval.
   - Press ESC to apply the changes.

13.8 Items Covered in This Tutorial

In this tutorial we have discussed the basic elements of data management in the program.

- Free format data files; interactive specification of data set descriptive information.
- BATS format data files.
- Saving and restoring model definitions.
- Choosing subsets of data for analysis.
- Appending new data to a data set.
- Aggregating time series.

13.9 Finale

If you have worked your way through the tutorials and reached this point confident of pursuing your own modelling, then we have achieved in some measure the goals we set for this book. If you have not, then the failure is ours and we would like to know of your difficulties so that we may address deficiencies in the book and the program in future editions. Whatever your reactions to the text and the software we would like to hear from you.

Part C

**BATS
REFERENCE**

Chapter 14

Communications

This part of the book is a reference guide to the BATS program. It is designed to assist you in your modelling with BATS once you have been through Parts A and B. The guide contains a formal description of every facility provided in the program, usage instructions, and communication details:

- Instructions for using menus, lists, dialogue boxes, tables, and interactive graphics.
- Menu option descriptions: details of every available option.
- File formats: details of every file type known to the program.
- Default Configurations

This chapter explains how to operate the main "interfaces" to BATS. We refer you to the tutorial sessions in Part B of the book for examples and step-by-step illustrations, the material here being restricted to a formal definition of procedures.

14.1 Menus

The following description applies to the DOS version of the program. Windows users are referred to the Windows documentation.

Menus are collections of options from which only one may be selected at any time. A menu is comprised of a title, list of alternative choices, and a summary help line. Menus are displayed at the top of the screen.

The menu option bar consists of space separated entries; on entry to a menu, the first, or leftmost, entry is highlighted or 'set off' from the remaining entries. The highlight denotes the current option. If you press the **ENTER** key the current option is selected. The option help bar displays a brief explanation of what the option will do for you.

The current option on the menu may be changed by using the left and right arrow cursor keys. (On some computers the keyboard has two sets of arrow keys: one dedicated set, and another set where the arrows are combined with numbers. The dedicated arrow keys will always move the current option; the combined numeric/arrow keys function as numerics when **Num Lock** is on—indicated by small light at the top of the keyboard—and as arrows when **Num Lock** is off. **Num Lock** is a toggle like the `Caps Shift` key: when it is on, pressing it once turns it off; and vice versa.)

The menu list is wraparound. Pressing the right arrow key repeatedly moves the current option successively to the right. When the last—the rightmost—option is reached, another right arrow key press will move the current option to the first entry—the leftmost—in the list; and vice versa left to right.

The option help information changes with the identified current option.

Selection

Option selection may be done in one of two ways.

(i) Make the required option current by using the arrow keys as described above, then press the **ENTER** key.

(ii) Press the alphabetic key corresponding to the first letter of the required entry (upper or lower case will do). If you press a key that does not correspond to any of the options, the keystroke will be ignored. If two or more options begin with the same sequence of letters you will have to enter additional keystrokes until a single option is uniquely identified. For example, on the Explore menu two options begin with the letter 'T': 'Time-plot' and 'Table'. Pressing 'T' (or 't') will reduce the menu to these two entries. Pressing 'i' (or 'I') will then choose 'Time-plot', while pressing 'a' (or 'A') will choose 'Table'. Pressing any other key following the 't' will abort the selection and return to the full Explore menu.

The last menu entry is always the 'quit' option which is selected when you have finished with the menu. On some menus this option is labelled 'Continue' or 'Return'. This occurs where a menu is presented in the middle of some 'larger' request and the term quit is reserved for that. 'Exit' on the root menu is another example.

The **ESC** key is a quick way of selecting the last menu entry.

Hot Key
Whenever BATS is in menu selection mode, pressing function key **F2** causes the current data and model summary to be displayed.

14.2 Lists

Lists are collections of items from which one or more may be selected. Lists are used for selecting data files and time series.

On starting a list the first entry in the leftmost column is highlighted as the current selection. The current selection is controlled by the following cursor movement keys.

- ↑ Make the entry in the row above the existing current selection (same column) the new current selection. If the existing current selection is in the first row, wrap around to the last row of the column.
- ↓ Make the entry in the row below the existing current selection (same column) the new current selection. If the existing current selection is in the last row, wrap around to the first row of the column.
- → Make the entry in the column to the right of the existing current selection (same row) the new current selection. If the existing current selection is the last column of the row, wrap around to the first column.
- ← Make the entry in the column to the left of the existing current selection (same row) the new current selection. If the existing current selection is the last column of the row, wrap around to the first column.
- **Page Up** Make the entry in the last row of the current column the new current selection.
- **Page Down** Make the entry in the first row of the current column the new current selection. (See also 'Special Selections' below.)

Making Selections
Use the **arrow/Page Up/Page Down** keys to make a required item the current entry (highlighted). Press the **ENTER** key. A marker appears to denote the selection. Pressing **ENTER** with a marked entry current removes the mark—the selection is cancelled. Mark all required entries, then press **ESC**.

Note that the selected entries are returned to the 'calling routine' in the order in which they appear in the list, *not* in the order in which you select them. This is important to bear in mind when selecting multiple series for cross-plots for example.

Some procedures—such as the model response series definition—allow only a single choice: as soon as you press **ENTER** the current entry is selected and the list is removed. Pressing the **ESC** key (after **ENTER**) is not necessary.

If ESC is pressed without making any selections, an empty selection list is returned to the calling function which takes appropriate action. For example, when reading a data file, no file is read if none is selected.

Special Selections

Some functions provide a default list of choices but allow you to specify an item of your own that is not included in the list. (This is only possible in single item selections.) Data file selection is an example. Page Down has a modified action in these cases: the list is removed and a dialogue box is displayed in its place. See the description of dialogue boxes below for usage details.

14.3 Tables

Tables display data or model estimate series in a square array format, one series per column, one time value per row. Data summary details are displayed at the top of the screen for reference.

Only a portion of a table is typically displayed at any one time because of the physical size limitations of the screen. The portion of the table on view may be changed using the following key strokes.

- ↑   Make the entry in the row above the existing current cell (same column) the new current cell. If the current cell is in the first row of the actual table, this key is ignored.

If the currently displayed portion of the table does not include the first row of the actual table and the current cell is in the first row of the displayed portion, then the portion of the table displayed is moved up by one row.

- ↓   Make the entry in the row below the existing current cell (same column) the new current cell. If the current cell is in the last row of the actual table, this key is ignored.

If the currently displayed portion of the table does not include the last row of the actual table and the current cell is in the last row of the displayed portion, then the portion of the table displayed is moved down by one row.

- →   Make the entry in the column to the right of the existing current cell (same row) the new current cell. If the current column is the rightmost column of the actual table this key is ignored.

If the currently displayed portion of the table does not include the rightmost column of the actual table and the current cell is in the rightmost displayed column, then the portion of the table displayed is moved right by one column.

14.3 Tables

- ← Make the entry in the column to the left of the existing current cell (same row) the new current cell. If the current column is the leftmost column of the actual table, this key is ignored.

If the currently displayed portion of the table does not include the leftmost column of the actual table and the current cell is in the leftmost displayed column, then the portion of the table displayed is moved left by one column.

- **Page Up** Moves the portion of the table currently displayed up by 25 rows. If fewer than 25 rows are available above the current display portion, then the first 25 rows of the actual table are displayed.

- **Page Down** Moves the portion of the table currently displayed down by 25 rows. If fewer than 25 rows are available below the current display portion, then the last 25 rows of the actual table are displayed.

- **Home** Make the cell in the top row, left column of the current display the current cell.

- **End** Make the cell in the bottom row, right column of the current display the current cell.

- **?** Invokes table display reformatting. The displayed portion of the table is removed from the screen and two dialogue boxes appear to the top right of the screen: 'cell (column) width' and 'decimal places'.

Following the rules for dialogue box entry (see below) a new display format is defined, for example, column width 15 with 6 decimal places for displayed numbers: Enter 15 and 6 in the two boxes respectively. **ESC** sets the entries and the table display is redrawn in the new format.

Note: If the column size is increased, then fewer columns may be displayed than there were to begin with. Such column truncation is performed from the right.

- **F** Write the table contents—not including row or column labels—to a named file. The file is written in the 'free' format (see Section 13.1 for details). File selection uses the list interface (described above in this section) with the currently configured directory and file mask for free format data files. The 'Special Selection' option (**Page Down** key) is available for 'arbitrary' file naming.

- **ESC** Exits the table display.

Editing

Table displays may appear in either 'view' or 'edit' mode, indicated by [View] or [Edit] above the first row label. In view mode the table contents may be inspected but not altered. Any keystroke other than those detailed above is ignored. In edit mode the table contents may be altered by typing numbers in standard form, such as '-1.6' (without the quotes), or in scientific notation, such as '-3.67e-2'. When you begin typing, the displayed contents of the current cell are erased and your input appears in its place. No change is made to the actual data value until the entry is completed by pressing the ENTER key.

Missing Values

Missing values in series are displayed as N/A. You may enter a missing value by typing either N or NA (lowercase is also acceptable) followed by pressing the ENTER key.

14.4 Dialogue Boxes

The program uses dialogue boxes whenever some 'general' user input is required, such as numerical values for component priors, or table formatting sizes for example.

Within a single box, nondestructive movement is possible using the right and left arrow keys. Movement between boxes in a multiple box display (a 'form') is achieved using the up and down arrow keys. The order in which the boxes of a form are traversed is dependent upon the form design. Movement is wraparound: the next position after the last position in a given box is the first position of the next box (which is the same box in a single box form) and vice versa. A short warning sound is emitted when the cursor is moved between boxes (including wraparound movements in a single box form).

Data Entry

Valid keystrokes for editing box contents are context-specific. For example, alpha keys are not allowed in numeric boxes as used for setting component priors for example. A warning sound is emitted when invalid keystrokes are entered.

Entry Complete

Press the ESC key when all box contents have been entered to have the program read the information.

Errors

Error handling in forms is context sensitive. On some occasions, such as the interactive specification of file details for free format file reading (see the FREE-format command on the Data/Input-Output/Read menu entry in Chapter 15), the program exits from the form and you may be asked if you wish to try again or to quit. On some occasions quitting is assumed.

On other occasions valid information *must* be given so the form remains displayed until an acceptable group of entries is made—for example, when setting component prior moments.

14.5 Graphs

In the DOS program all graphs, with the exception of the dynamically drawn on-line and retrospective graphs, may be customised using the interactive graphics options detailed here. Windows users may use the Windows native facilities to cut and paste BATS displays into other applications, such as the paint program, for editing and printing.

Mouse

If your computer has a mouse device driver loaded and a mouse device connected to the system (and BATS recognises your mouse), then the mouse cursor—usually an arrow—appears on the upper left of the display when interactive graphics is first entered. The mouse cursor responds to movements in the mouse device and to certain keyboard key presses.

Cursor

If your computer is not equipped with a mouse pointing device—or BATS cannot detect your mouse—then a cross-hair cursor is used. This cursor responds to the same movement keystrokes as the mouse. (See also the comments on keyboard arrow keys in the menus description above.)

- ↑   The cursor is moved one drawing unit (or 'pixel') upwards.
- *Shift* ↑   The cursor is moved 15 drawing units upwards.
- ↓   The cursor is moved one drawing unit downwards.
- *Shift* ↓   The cursor is moved 15 drawing units downwards.
- →   The cursor is moved one drawing unit to the right.
- *Shift* →   The cursor is moved 15 drawing units to the right.
- ←   The cursor is moved one drawing unit to the left.
- *Shift* ←   The cursor is moved 15 drawing units to the left.

Annotations

The graphical annotation facilities are selected by single key press from the following list. *Note that interactively added annotations cannot be removed once they have been created.*

- a   Draw an arrow in the current colour (**k**) and line style (**l**) from the last marked point (**m**) to the current cursor location.

- c   Draw a line in the current colour and line style from the last marked point to the current cursor location; extend the line (in both directions) to the time frame border.
- d   Draw a line in the current colour and line style from the last marked point to the current cursor location.
- k   Change the current drawing colour. A window pops up listing the available colours. Press the number corresponding to the colour you wish to use for subsequent drawing.
- l   Change the current line drawing style. A window pops up listing the available styles. Press the number corresponding the the style you wish to use for subsequent drawing.
- m   Mark the current cursor location as the last marked point.
- r   Move the cursor to the last marked position (m).
- t   Add text in the current colour. The location cursor (mouse or cross-hair) is removed and replaced by a double bar cursor. Any text subsequently typed will appear on the screen, until the **ENTER** key is pressed to terminate text input. The **backspace** key may be used to erase text one character at a time *during the initial input phase only* (see the warning at the start of 'Annotations').
- x   Draw a symbol in the current colour and line style at the current cursor location. A window pops up listing the available symbols. Press the number corresponding to the symbol you wish to draw.
- w   Where am I? An information box appears containing the x-y location of the cursor. Note that the x coordinate (representing time in most of BATS' graphs) is displayed as an integer and not as a cycle/time pair.

Printing

Printing graphical displays *cannot* be done directly from within BATS. You can, however, store PostScript descriptions of graphs in files and subsequently print the files from DOS. Pressing **s** in interactive graphics mode enables you to store a PostScript description of the current display on file. A window pops up in which to enter the file name. The file name may include a full path definition. If it does not, the file will be written in the configured PostScript directory (see the **Directories** command on the Configuration menu entry in Chapter 15 for file configuration details). The PostScript files may be edited and included in other documents: that's how we created the graphs for this book. [Windows users should read the note at the beginning of this section.]

Chapter 15

Menu Descriptions

This chapter details each of the menus in the BATS program. Every menu option is described. We begin with a schematic of the entire menu structure, then detail the menus in order reading down this representation. Note that in action BATS suppresses options that are not meaningful for a given state of data, model definition, or analysis, so that when running the program a particular menu may not display all of the options shown here.

15.1 Menu Trees

The menu control structure in BATS is hierarchical with a single starting point conventionally referred to as the 'root'. The hierarchy is depicted in the next subsection. Several singlelevel menus may appear during the course of an analysis depending upon the state of the analysis. These 'floating' menus are depicted in the subsection following the main hierarchy.

15.1.1 Main Menu Tree

Data
    Input/Output
        Read
            BATS format
            FREE format
            Read Model

               Quit
           Write
               BATS format
               FREE format
               Read Model
               Quit
           Keyboard
           Aggregate
           Create
           Delete
           Select
           Append
           Edit
           Quit
   Explore
           Time-Plot
           MPlot
           H'Gram
           QQPlot
           P'Gram
           Autocor
           X-Y
           Table
           Corr
           Seas
               On-line
               Retrospective
               Both
               Quit
           Factor
           Quit
       Clear
       Quit
   Reset
       All
       Outliers
       Interventions
       Discounts
       Quit
   Model
       Components
           Response

15.1 Menu Trees

    Trend
      Constant
      Linear
      Quit
    Seasonal
      Free-form
      Restricted
      Quit
    Regression
    Variance
    Quit
   Interrupts
    Outlier-Intervention
    Monitor
    Quit
   Discount
    Constant
    Individual
    Quit
   Remove
   Quit
Fit
   Reference
   New-Prior
   Existing-Prior
    Trend
    Seasonal
      Exchangeable
      Constant Variance
      Full
      Quit
    Regression
    Variance
   Quit
Retrospective
Predict
   View-Priors
   Edit-Priors
   Abort
   Continue
Configuration
   Graphics

　　　　　Auto-Detect
　　　　　AT&T
　　　　　CGA
　　　　　EGA
　　　　　Hercules
　　　　　VGA
　　　　　Quit
　　　Directories
　　　Monitor
　　　Store
　　　DOS
　　　Estimates
　　　Quit

15.1.2 Floating Menus

The following menus occur in several places in BATS. The forecast priors menu occurs in intervention and prediction; the monitor signal menu appears on the forecast graph each time the monitor signals during an analysis; the forward intervention menu appears each time intervention is selected during an analysis (either in response to a monitor signal, option 3, or at prespecified intervention times).

Forecast Priors
View-Priors
Edit-Priors
Abort
Continue

Monitor Signal
1 Ignore
2 Automatic
3 User

Forward Intervention
View-Priors
Edit-Priors
Interrupts
Predict
Abort
Return

15.2 Root

The Root menu is the starting point of a BATS session, after the initial screen logo is displayed. It is also the place from which each of the main activities—data management, model specification, and analysis—are accessed. Each of these elements are menus in their own right, with various levels of submenus below them. The entire menu structure is arranged in a hierarchical form with the Root serving as the focus. The complete hierarchy is detailed below.

- **Data** The Data command provides access to the program's data management facilities, split into two groups of operations: basic data access and editing, and exploration. Each group is accessed as a further level in the menu hierarchy, detailed in the Data menu and its component menus below.

- **Reset** The Reset command enables quick and easy reinitialisation of analysis operating conditions. Specifically, options are available for restoring all model component discount factors to the current user defined routine values; removing all existing outlier indicators; and removing all existing forward intervention points. These operations may be performed as a group or individually as described in the entry for Reset below.

For analyses that include the use of the forecast monitor or user-specified forward interventions this command is useful for removing the effects of past analyses. For simple analyses which do not use monitoring or intervention this command will not be used.

- **Model** The Model command provides access to the program's model definition facilities and analysis environment conditions. Individual model component structures, namely response series, trend, seasonal, regression, and observation variance, and analysis operating conditions, namely component discount factors, forecast monitoring, and forward interventions, are accessed through entries on the model submenus, detailed in Model and its component menus below.

- **Fit** The Fit command initiates the on-line analysis of the current data set using the defined model under the specified operating conditions just described for the Model command. First though, one further set of operating conditions must be determined—the component prior values—detailed in Fit and its component menus below. Once the priors are set, the standard time frame is displayed and the forward filter analysis proper is begun. During the analysis the component estimates are stored in the file ONLINE.RES for later use. When the on-line analysis is complete, individual component estimates (and uncertainty intervals) are computed which may subsequently be examined using the Data/Explore menu commands. The on-line forecast graph remains displayed until a keystroke is entered, whereupon the program returns to the root menu. There are no interactive graphical facilities available with this forecast graph.

- **Retrospective** The **Retrospective** command launches immediately into the back filtering analysis corresponding to the most recent on-line analysis. The on-line state parameter moments are read from the storage file ONLINE.RES and retrospective results stored automatically in file SMOOTH.RES for later use. When complete the fitted values graph remains displayed until a keystroke is entered, whereupon the program returns to the Root menu. There are no interactive graphical facilities available with the retrospective graph.

- **Predict** The **Predict** command computes model based predictions using subjectively specified prior moments for model components. Prediction may be performed as soon as a model definition is set (which requires at least the response series and a trend component defined) in which case predictions are computed from the starting point of the data; or following an analysis (on-line only or on-line and retrospective) in which case predictions are computed from the first time subsequent to the data.

If no analysis has yet been performed, the **Predict** command will immediately display the component prior entry forms for each of the defined model components, followed by another form to set the forecast horizon (last forecast time). Finally the Forecast Priors menu appears, allowing the priors just set to be examined or altered before continuing with the calculations, or the prediction exercise to be aborted.

- **Configuration** The **Configuration** command provides access to the facilities for tuning the operating environment of the program. The available customisations are accessed from a further level in the menu hierarchy, detailed in Configuration and its component menus below.

- **Exit** The **Exit** command terminates your BATS session. Before doing so, you are asked for confirmation.

15.3 Data

The Data menu (displayed by the Data command from the Root menu) provides access to the program's data management facilities. These facilities are divided into two groups, one dealing with data entry, data modification, and data exit; and a second dealing with data examination and exploration. Groups are accessed by submenu, each of which is detailed below.

- **Input-Output** The **Input-Output** command provides access to the program's data entry, modification, and output facilities. Data sets are made known to BATS in the first instance by reading a data file from disc. Once entered, data may be edited within the program and subsequently written to file once more as required. Individual options are detailed in the Data/Input-Output menu and its component submenu descriptions below.

- **Explore** The `Explore` command provides access to all of the data series and model estimates from the most recent analysis. A range of display and data analytic facilities is available for examining these series, from time plots through specialised graphs to tables. Individual options are detailed in the Data/Explore menu below.

- **Clear** The `Clear` command erases the current data set and model definition.

- **Quit** The `Quit` command returns you to the Root menu.

15.4 Data/Input-Output

The Data/Input-Output menu provides access to the program's data entry, modification, and output facilities. Data sets are made known to BATS in the first instance by reading a data file from disc. Once entered, data may be edited within the program and subsequently written to file once more as required. Note that data sets modified within the program are not automatically saved to file: explicit instruction is required.

- **Read** The `Read` command provides menu access to the data file input facilities. File format options are detailed in the Data/Input-Output/Read menu below.

- **Write** The `Write` command provides menu access to the data file output facilities. File formats are detailed in the Data/Input-Output/Write menu below.

- **Aggregate** The `Aggregate` command allows a data series to be cumulated in blocks. For example, if a data set is currently defined as monthly with twelve values per year, and it is desired to analyse the data on a quarterly basis, the series must be cumulated in blocks of three months. If there are missing values in the original series, the blocks of which they are elements are defined as missing. Continuing with the monthly to quarterly example, if the original data set begins at February, then February and March should be dropped from the quarterly series which will begin with quarter two (comprising the sum of months April, May, and June). This arrangement is not enforced, and you may begin a 'quarterly' cumulation in February if you wish. All the data descriptive details are entered in a form with the usual dialogue box operation. Chapter 13 walks you through an illustration. Chapter 14 details dialogue box usage.

If an analysis exists for the originally scaled data, the program will warn you and you may either delete the analysis results or abort the data aggregation.

- **Create** The `Create` command enables new series to be defined from transformations and combinations of existing series. Any existing series in a data set may be transformed by the functions listed below (Chapter 10 walks you through an illustration). Missing values are always transformed to missing values. Invalid values for a transformation, square root of a negative for example, are also transformed to missing values.

Two dialogue boxes are displayed. One for the 'target series', that is, the name of the series to be created or overwritten; and one for the transformation formula. The formula will be in one of three forms according to the type of transformation as the examples below illustrate.

Example formula 1: `sqrt SALES`

- o `log10` Decimal logarithmic transformation of the source series.
- o `loge` Natural logarithmic transformation of the source series.
- o `sqrt` Square root transformation of the source series.
- o `exp` Exponential transformation of the source series.
- o `center/centre` Subtracts the series mean from each series value.
- o `std` Centres the series then divides by the series standard deviation.
- o `power` Raises the source series to the specified power index.

Example formula 2: `2 ma SALES`

- o `ma` Centred moving average: `2 ma SALES` calculates a five point (two either side) centred moving average. End points are set to missing values.
- o `mm` As for the moving average, but a moving median.
- o `ewma` Exponentially weighted moving average.
- o `lag` Series lagging operator.
- o `diff` Series differencing operator.

Example formula 3: `SALES + 3` or `SALES + PRICE`

- o `+` Sum two series, or add a constant to a series.
- o `−` Subtract two series, or subtract a constant from a series.
- o `×` Multiply two series, or multiply a series by a constant.
- o `/` Divide two series, or divide a series by a constant.

- **Select** The `Select` command allows a subset of the data to be selected for analysis: a portion of the data at the beginning of the existing set may be discarded by setting a new data start time; and a portion at the end of the existing set may be discarded by setting a new data end time.

- **Edit** The `Edit` command invokes the program data editor. This is described in detail as Tables in Chapter 14.

- **Quit** The `Quit` command returns you to the `Data` menu.

15.5 Data/Input-Output/Read

The Data/Input-Output/Read menu provides data file input from three possible file formats. This is the only mechanism by which data may be brought into the program.

- **BATS-Format** The BATS-format command instructs the program to display a list of all of the files in the BATS format data directory that have the configured file name extension. (See the Directories command on the Configuration menu below for details of setting default file directories and names.) As described in the operation of lists in Chapter 14 the Page Down key may be used to open up a dialogue box for entering a file name that does not appear in the default list. Files in the BATS format contain data descriptive information in addition to the raw series values. The data descriptive details in the selected file are loaded into BATS, any existing data (and associated model and analysis if any) being erased first.

- **FREE-Format** The FREE-format command instructs the program to display a list of all of the files in the free format data directory that have the configured file name extension. (See the Directories command on the Configuration menu below for details of setting default file directories and names.) As described in the operation of lists in Chapter 14 the Page Down key may be used to open up a dialogue box for entering a file name that does not appear in the default list. Files in the free format contain data series values, one series per column and one time point per row, and no other information. Series names and data timing information are entered interactively after the file contents are loaded. Chapter 13 walks you through an illustration of how to load a free format data file and explains what descriptive information is subsequently entered. The data descriptive details in the selected file are loaded into BATS, any existing data (and associated model and analysis if any) being erased first.

- **Read-Model** The Read-Model command reads a data file in the BATS format as described above, and in addition loads a model definition if one is found appended to the series values (see the Write-Model command on the Data/Input-Output/Write menu below). If no model definition is recorded in the selected file, this command performs exactly as the BATS-format command above.

- **Quit** The Quit command returns you to the Data/Input-Output menu.

15.6 Data/Input-Output/Write

The Data/Input-Output/Write menu provides data file output in the same three file formats as used in the 'read' operation.

- **BATS-Format** The `BATS-format` command instructs the program to display a list of all of the files in the BATS format data directory that have the configured file name extension. (See the Directories command on the Configuration menu below for details of setting default file directories and names.) As described in the operation of lists in Chapter 14 the Page Down key may be used to open up a dialogue box for entering a file name that does not appear in the default list. The currently loaded data set is written to the selected file.

- **FREE-Format** The `FREE-format` command instructs the program to display a list of all of the files in the free format data directory that have the configured file name extension. (See the Directories command on the Configuration menu below for details of setting default file directories and names.) As described in the operation of lists in Chapter 14 the Page Down key may be used to open up a dialogue box for entering a file name that does not appear in the default list. The currently loaded data set is written to the selected file.

- **Write-Model** The `Write-Model` command instructs the program to write the currently loaded data set to file using the BATS format as described above; the current model definition is then appended.

- **Quit** The Quit command returns you to the Data/Input-Output menu.

15.7 Data/Explore

The Explore menu provides access to all of the data series and model estimates from the most recent analysis. A range of display and data analytic facilities is available for examining these series, from time plots through specialised graphs to tables. Series analysis is a two stage procedure. First the type of display is selected, then the desired series are selected from the usual list display. Multiple series may be selected for analysis; the specified analysis procedure will take appropriate action for each series so selected. For example, the `Time-Plot` command will display a series of individual time plots, one for each selected series.

Note that selected series are presented to the analysis functions in the order in which they appear in the list, not in the order in which you select them.

Estimate series names are prefixed with O for on-line or S for smoothed (retrospective). In addition, some estimate series are prefixed with * which indicates that the response series will be graphed along with the estimates. The estimate series names are as follows. Each of the O names has a corresponding S series wherein 'on-line' is replaced by 'retrospective'. These names have been omitted in the interest of brevity.

- *O Level On-line estimated level.

15.7 Data/Explore

- O Growth   On-line estimated growth (if the trend component is linear).
- O Seasonal   On-line seasonal estimate.
- O Factor n On-line n^{th} seasonal factor, where n indexes from 1 to the seasonal period—twelve for monthly data for example. (Seasonal factor estimates may be 'turned off', that is, not computed even when a seasonal component is included in the model. See the **Estimates** command on the **Configuration** menu below.)
- O PRICE coeff   On-line regression coefficient for regression variable PRICE. In practice the names of the regressor variables in your model will appear.
- O PRICE effect   On-line regression effect (coefficient times regressor) for regression variable PRICE. In practice the names of the regressor variables in your model will appear.
- O Regression Block n On-line overall regression block effect. This estimate only appears if more than one regressor is included in the model. At present only a single regression block may be defined so n will always be 1. (Regression block estimates may be 'turned off', that is, not computed even when there are several regressors in a block. See the **Estimates** command on the **Configuration** menu below.)
- O Std Dev   On-line estimated observation standard deviation.
- *Forecasts   One step ahead forecasts. The response series will be included in time plots.
- *Online Fit   On-line estimated fitted values. The response series will be included in time plots.
- O Raw Residuals   One step ahead forecast residuals.
- O Std Residuals   One step ahead standardised forecast residuals.
- *Fitted   Retrospective estimated values.

The Data/Explore command options are:

- **Time-Plot**   The **Time-Plot** command displays a single series time plot. Data series are displayed as small circles at the data values joined by a line. Model estimate series are displayed as a mean value line with dashed lines showing the 90% probability limits. When multiple series are selected for display, each series is displayed in turn: quitting the first graph automatically brings up the second and so on.
- **MPlot**   The **MPlot** command displays one, two, or three series in a single graph. Each series is displayed in a separate frame so that the full screen is partitioned into (up to) three segments. Data and estimate series are drawn in the same format as in the **Time-Plot** command. When more than three series are selected, **MPlot** takes the series in batches of three:

quitting the first graph (the first three series) automatically brings up the second graph and so on.

- **QQPlot** The `QQPlot` command displays a normal probability quantile plot of a series. This is useful for examining (standardised) model residuals to check for departures from the normal distribution assumption.
- **P'gram** The `P'gram` command displays a periodogram analysis of a series. The periodogram is useful for identifying cycle periods in stationary series. It is less useful with nonstationary series.
- **Autocor** The `Autocor` command displays the first 10 sample autocorrelations of a series. This is useful for detecting residual structure in estimated model errors.
- **X-Y** The `X-Y` command displays a conventional cross-plot of two series. If only one series is selected, the command is ignored. If more than two series are selected, the series are taken in batches of two: quitting the first graph automatically brings up the second graph and so on. If an odd number of series is selected, say three, then the last series is ignored.

Note that the X-Y ordering of series is determined by the positions of the names in the list. It is not possible to reverse the X-Y ordering.

- **Table** The `Table` command displays selected series using the table interface (in view mode) described in Chapter 14. Note that estimate series take up two columns: one column for the mean estimate, and one for the 90% deviation from the mean. The observation variance estimate appears as a mean value and the associated degrees of freedom.
- **Corr** The `Corr` command displays the posterior correlation matrix for the model parameters as computed at the end of the analysis. The display uses the table format in view mode. Note that when a seasonal component is included in the model definition, correlations are given for the factors, not the harmonic components. The correlation table appears directly the `Corr` command is selected: no list display for series selection is presented.
- **Seas** The `Seas` command displays estimated seasonal factors in a two frame graph. The upper frame shows the factors as perturbations from the underlying estimated series level (trend plus any included regressions); the lower frame shows just the factor estimates. Neither frame gives 90% uncertainty limits. If the retrospective analysis has been performed, the `Seas` command brings up a menu (see **Data/Explore/Seas** below) for choosing among the on-line and retrospective estimates. If only an on-line analysis is available, no such menu appears and the on-line estimates are displayed immediately. No list display for series selection is presented by this command.
- **Factor** The `Factor` command displays estimated seasonal factors in a multiframe graph. The series list display is restricted to the seasonal fac-

tors, and up to six on a graph will be drawn. For each factor, the temporal development of the factor estimate is shown (without uncertainty limits). This display is useful in determining where major changes in seasonal patterns take place, and precisely how the patterns change.

- **Quit** The Quit command returns you to the Data menu.

15.8 Data/Explore/Seas

The Seas menu is displayed by the Seas command on the Data/Explore menu only when there are both on-line and retrospective estimates available. The options allow choice of which estimates to display. See the description of the Seas command on the Data/Explore menu above for a details of the form of the graph.

- **On-Line** The On-Line command instructs the program to display the seasonal factor plot for the on-line estimates only.

- **Retrospective** The Retrospective command instructs the program to display the seasonal factor plot for the retrospective estimates only.

- **Both** The Both command instructs the program to display the seasonal factor plot for both the on-line and the retrospective estimates. The on-line estimates are displayed first; the retrospective estimates are displayed after you quit from the on-line graph.

- **Quit** The Quit command instructs the program to abort the seasonal factor plot: no graphs are drawn, and the program returns you to the Data/Explore menu.

15.9 Reset

The Reset menu (displayed by the Reset command from the Root menu) provides options for quick and easy reinitialisation of analysis operating conditions. Specifically, options are available for restoring all model component discount factors to the current user-defined routine values, removing all existing outlier indicators, and removing all existing forward interventions points. For analyses that include the use of the forecast monitor or user-specified forward interventions, this command is useful for removing the effects of past analyses. For simple analyses which do not use monitoring or intervention, this command will not be used.

- **All** The All command combines the operations of the Outliers, Interventions, and Discounts commands on this menu. That is, all existing outlier and intervention indicators are removed, and all discount factors are set to their component routine values.

- **Outliers** The `Outliers` command removes all existing marked outlier indicators. Outliers are set when monitor signals are responded to with the automatic rules *if the monitor signal has run length one*. Such outliers should typically be removed in subsequent analysis. But note that user specified outliers are not distinguished from those set by monitor response: all are removed by this command. Individual outliers may be set or removed using the `Intervention-Outlier` command on the Model/Interrupts menu.
- **Interventions** The `Interventions` command removes all existing forward intervention points. Forward interventions are set when monitor signals are responded to with subjective user intervention. Note, however, that user-specified forward interventions are not distinguished from those set by monitor response: all are removed by this command. Leaving intervention points set does not alter a subsequent analysis; it merely interrupts the analysis at the set times, allowing you to intervene or not as you choose.
- **Discounts** The `Discounts` command resets component discount factors for all times to the routine component discount values. If forecast monitoring was used in a previous analysis, then any signals responded to using the automatic adjustment rule will have set the component discounts for such times to the exception values (see the `Monitor` command on the Configuration menu). These discounts will typically not be desired (by default) in a subsequent analysis.
- **Quit** The `Quit` command returns you to the Root menu.

15.10 Model

The Model menu (displayed by the `Model` command from the Root menu) provides access to the program's facilities for model definition and analysis environment specification. There are options for setting model components and routine discount factors, outlier and forward intervention points, and forecast monitors.

- **Components** The `Components` command provides access to the menu of model components for model definition. Individual components are detailed in the Model/Components menu and its submenus below.
- **Interrupts** The `Interrupts` command provides access to the menu analysis environment for setting outliers, forward interventions, and forecast monitors. Details are given in the Model/Interrupts menu below.
- **Discount** The `Discount` command gives access to alternative modes of component discount factor specification. Details in the Model/Discount menu below.
- **Remove** The `Remove` command deletes the current model definition—including resetting all forecast monitors to 'off', and removing all existing

15.12 Model/Components/Trend   **395**

outliers and forward interventions—and returns you to the Root menu.

- Quit   The Quit command returns you to the Root menu.

15.11 Model/Components

A model requires both a response series and a trend component to be defined. The other components, seasonal, regression, and variance (power law), are optional. When a component is defined its discount factors for each time are set to the component's default value. Therefore, if you wish to alter component discount factors, you should do so after the component form has been decided. For example, if you set a constant trend, then change the routine discount factors from the program default value 0.9 to, say, 0.95, a subsequent change of mind and alteration of the component to a linear form, will automatically reset the trend discounts to 0.9.

The current model definition may be displayed at any time by pressing the function key F2.

- Response   The Response command allows you to select the model response series. A list of the names of all of the series that are not currently included in the model as regressors is displayed from which to choose the response. List selection is detailed in Chapter 14.

- Trend   The Trend command allows you to select the form of the trend component for your model. The available forms are displayed as the Model/Components/Trend menu which is detailed below. A trend component (in addition to the response series) *must* be specified before BATS regards a model as defined.

- Seasonal   The Seasonal command allows you to select the form of a seasonal component for your model. The available forms are displayed as the Model/Components/Seasonal menu which is detailed below.

- Regression   The Regression command allows you to select regressors for a regression component. A list of the names of all of the series except the currently defined response series is displayed from which to choose. Multiple series may be selected as regressors. Selecting no series means that no regression component is defined. List selection is detailed in Chapter 14.

- Variance   The Variance command allows you to specify a power law for the observation variance. A dialogue box is displayed, initially containing the current power law setting (zero is the standard default, meaning no power law is in effect). Dialogue box usage is detailed in Chapter 14.

- Quit   The Quit command returns you to the Model menu.

15.12 Model/Components/Trend

The Model/Components/Trend menu allows you to select the form of the trend component for your model. A trend component (in addition to the response series) *must* be specified before BATS regards a model as defined.

- **Constant** The Constant command defines a dynamic constant trend component. The component system equation is

$$\text{Level}_t = \text{Level}_{t-1} + \omega_t.$$

- **Linear** The Linear command defines a dynamic straight line trend component. The component system equations are,

$$\text{Level}_t = \text{Level}_{t-1} + \text{Growth}_{t-1} + \omega_{1t},$$
$$\text{Growth}_t = \text{Growth}_{t-1} + \omega_{2t}.$$

- **Quit** The Quit command returns control to the Model/Components menu.

15.13 Model/Components/Seasonal

The Model/Components/Seasonal menu allows you to select the form of a seasonal component for your model.

- **Free-form** The Free-form command selects an unrestricted seasonal component. An arbitrary pattern of seasonal variation may be modelled with such a component as it provides a separate parameter—a seasonal factor—for each time interval. For example, in a data set with monthly observations (and period one year) there will be one seasonal factor for each month: twelve in all. Of course, seasonal factors are constrained to sum to one over a full year because a seasonal component models seasonal deviations from an underlying level. It is the responsibility of the trend and regression components to model this underlying level.

- **Restricted** The Restricted command allows you to specify a constrained pattern of seasonal variation. Seasonal components are modelled as combinations of harmonics, and constraints are implied by including only a subset of these harmonics. (The unrestricted model set by the Free-form option is equivalent to selecting the full set of harmonics.) Two dialogue boxes are presented for this specification: one for the period of the seasonality (in case you wish to use a period different from the data set period—a rather unusual circumstance), and one for the required harmonic indices.

- **Quit** The Quit command returns control to the Model/Components menu.

15.14 Model/Interrupts

The Model/Interrupts menu provides access to facilities for tuning the analysis environment for the current model definition. Response series values may be preset as 'outliers', analysis suspension times may be preset, and forecast monitors may be switched on.

- **Outlier-Intervention** The Outlier-Intervention command allows you to set response series values to be ignored in subsequent analysis by designating them as 'outliers', and to set times for interruption of analyses (to calculate predictions or perform subjective intervention for example). Outlier and intervention times are set using the table mechanism with interventions in the first column and outliers in the second column. A zero entry for a given time (the default) indicates *no* intervention/outlier; a nonzero entry indicates an intervention/outlier. The operational details of the table mechanism are described in Chapter 14.

- **Monitor** The Monitor command allows forecast monitoring to be switched on or off. BATS provides three forecast monitors: one for detecting upward changes in level, one for detecting downward changes in level, and a catchall for detecting nonspecific changes. A form comprising three dialogue boxes, one for each monitor, is displayed. Enter N for a monitor that is not required, Y for a monitor that is required. Dialogue box usage is detailed in Chapter 14.

Chapter 3 gives formal details of forecast monitoring. Chapter 11 describes the operational mechanism and works through a detailed example. See also the entry for Monitor command on the Configuration menu in this Chapter.

- **Quit** The Quit command returns you to the Model menu.

15.15 Model/Discount

The Model/Discount menu provides access to the discount factors for each of the model components currently defined. Discount factors may be individually set for each time.

- **Constant** The Constant command is a quick way of assigning a single set of component discount factors for all times. A series of dialogue boxes appears, one for each defined model component, initially containing those components' routine discount factor values. The value entered in the trend box is set as the trend component discount for all times; and similarly for any other defined components. Warning: if you set nonroutine discount factors for some times (see the Individual command next) subsequent use of this command will override such values. Dialogue box usage is detailed in Chapter 14.

- **Individual** The Individual command gives you access to the component discount factors for all times using the table format (in editing mode to allow changes). Each defined model component has a column in the table. Table usage is detailed in Chapter 14.

- **Quit** The Quit command returns you to the Model menu.

15.16 Fit

The Fit menu (displayed by the Fit command from the Root menu) instigates the on-line analysis of the currently defined model. Component priors have to be set first.

- **Reference** The Reference command instructs the program to initialise the on-line analysis with default non-informative priors. An approximation to the formal reference analysis is used in the program so that predictions may be made from the first time point. (In the formal case, prediction is not possible until several data points have been analysed and sufficient information extracted to define proper parametric distributions.)

- **New-Prior** The New-Prior command instructs the program that you wish to specify component priors.

- **Existing-Prior** The Existing-Prior command instructs the program to use the component priors that were used to initialise the previous analysis. But before the analysis is actually initiated you are given the opportunity to examine or change any of the priors—see the Fit/Existing-Prior menu below.

- **Quit** The Quit command terminates the on-line analysis request and returns you to the Root menu.

15.17 Fit/Existing-Prior

The Fit/Existing-Prior menu allows you to modify component priors before initiating an on-line analysis, having copied the priors from the most recent previous analysis (see the Fit menu above).

- **Trend** The Trend command provides the means by which the trend component prior specification may be examined and altered. A form is displayed comprising dialogue boxes for the prior mean and standard deviation for each of the trend component parameters. For a constant trend component there is just the level parameter; for a linear trend component there are the level and growth parameters. Dialogue box usage is detailed in Chapter 14.

- **Seasonal** The Seasonal command provides the means by which the seasonal component prior specification may be examined and altered. A seasonal prior may be specified in several ways, the alternatives being accessed from a submenu—see Fit/Existing-Prior/Seasonal below.

- **Regression** The Regression command provides the means by which the regression component prior specification may be examined and altered. A form is displayed comprising dialogue boxes for the prior mean and standard deviation for the coefficient of each regression variable included in the model. Dialogue box usage is detailed in Chapter 14.

- **Variance** The Variance command provides the means by which the observation variance component prior specification may be examined and altered. A form is displayed comprising dialogue boxes for the observation standard deviation estimate and the associated degrees of freedom.

- **Quit** The Quit command signals that you have completed your prior specification and the program gets on with the on-line forecasting.

15.18 Fit/Existing-Prior/Seasonal

The Fit/Existing-Prior/Seasonal menu provides alternative ways to examine and specify the seasonal component prior.

- **Exchangeable** The Exchangeable command sets an exchangeable seasonal prior. All seasonal factors have prior mean zero, and a common factor standard deviation is specified in a dialogue box. Dialogue box usage is detailed in Chapter 14.

- **Constant-Variance** The Constant-Variance command sets a seasonal prior with individually specified factor means and a common factor variance. A form is displayed comprising dialogue boxes for the factor means and the common factor standard deviations. The program automatically constrains the factor means to have zero sum. Dialogue box usage is detailed in Chapter 14.

- **Full** The Full command sets a seasonal prior from complete user specification of individual factor means and standard deviations. A form is displayed comprising dialogue boxes for each of the factor means and standard deviations. The program automatically constrains the means to sum to zero, and appropriately constructs the factor covariance matrix to reflect this constraint and the specified factor standard deviations. Dialogue box usage is detailed in Chapter 14.

- **Quit** The Quit command returns you to the Fit/Existing-Prior menu.

15.19 Forecast Priors

The Forecast Priors menu (displayed by the **Predict** command from the Root menu or the Forward Intervention menu) provides access to the model component prior specifications for examination and alteration. The prediction operation may also be cancelled from this menu.

- **View-Priors** The **View Priors** command presents each model component prior in turn for viewing. The parameter means and standard deviations are displayed for each defined component, one component at a time. (The observation variance is displayed as the mean and degrees of freedom.) A component display remains on screen until a key stroke is entered, whereupon the next component display appears. None of the model prior moments may be altered with this command.

- **Edit-Priors** See the **Existing-Prior** command on the Fit menu.

- **Abort** The **Abort** command instructs the program to cancel the prediction and return you to the Root menu.

- **Continue** The **Continue** command instructs the program to take the component prior specifications and go on with the prediction.

15.20 Configuration

The Configuration menu (displayed by the **Configuration** command from the Root menu) provides a set of options for tuning the operating environment of the program. The following sections detail the available customisations.

- **Graphics** The **Graphics** command provides a means of selecting alternative graphics modes (DOS version only). BATS automatically detects the highest possible graphical display resolution available on your computer. Ordinarily this will be the most suitable mode for running the program and the **Graphics** command will not be used. However, should you need, or wish, to set an alternative display mode, you may do so here. Support is provided for five popular graphics modes: AT&T, IBM Colour Graphics Adaptor (CGA), Enhanced Graphics Adaptor (EGA), Video Graphics Array (VGA), and Hercules monochrome. If your computer does not have one of these (or compatible) graphics display adaptors BATS will probably not run.

WARNING: *Attempting to run a program in a video mode not supported by your computer can result in physical damage to the video display system.*

- **Directories** The **Directories** command allows you to tell BATS where to look for your data files (Directory) and how to recognise your file names (File Mask). You can also define where PostScript (graph) output

15.20 Configuration

files will be created and where BATS is to store its session work files. A form comprising a set of seven dialogue boxes arranged in two columns is displayed; the columns are labelled 'Directory' and 'File Mask'.

When the program is run for the first time all of the directory entries default to the session startup directory. File names are assumed to have extensions '.DAT' for BATS format data files and '.FRE' for free format data files. For example, with this configuration BATS will display a sorted list of all those files in the startup directory that have file name extension '.DAT' when you select the **BATS-format** command from the **Data/Input-Output/Read** or **Data/Input-Output/Write** menus. The appendix to Chapter 9 gives a detailed description.

- **Monitor** The **Monitor** command allows you to determine the nature of the forecast monitors—basic monitor sensitivities and cumulative run length boundary—and the component exception discount factors.

For each of the three monitors—scale inflation, level increase, and level decrease—monitor sensitivity is determined by the (forecast density of) the alternative model, and a threshold value for deciding by how much the alternative forecast must be 'better than' the standard forecast before a signal is generated. For example, in the level increase monitor, the alternative model, to which the model forecast is compared by the monitor procedure, is defined to have the same general forecast distribution as the standard model but with a mean increased by n standard deviations. The quantity n is all that is required to fix the alternative model. Model forecasts are compared using the Bayes factor, for which a signal threshold is specified. The exception discount factors are used to compute model component priors when automatic signal response is selected for a monitor signal of run length one. Chapter 3 gives formal details of forecast monitoring. Chapter 11 describes the operational mechanism and works through an example. See also the **Monitor** command on the **Model/Interrupts** menu above.

A form comprising a set of 11 dialogue boxes is displayed in which the monitor and exception discount factor information is entered. Dialogue box usage is detailed in Chapter 14.

- **Store** The **Store** command copies the current configuration information for files and directories into a file called BATS.CFG located in the session startup directory. You may notice that the status line displays the message "Saving configuration file" while the file is being written.

Any subsequent BATS session initiated from the same directory will automatically load these configuration settings at startup time. Note that the configuration file *must* be located in the startup directory in order for BATS to find it. If no such file is found, the default configuration is set.

- **DOS** The **DOS** command allows the BATS session to be temporarily interrupted and control returned to the computer's operating system. BATS remains resident in the computer's memory so that the temporary DOS session will have implicit size limitations on the tasks that can be performed. To return to BATS from the temporary DOS session type **exit** at the DOS prompt and press **ENTER**. BATS will reappear with the Configuration menu displayed. For obvious reasons this option is not included in the Windows version of the program.

• **Estimates** The **Estimates** command is used for telling BATS which of a set of optional component estimates should be routinely calculated following an analysis. The optional estimates are the individual seasonal factors, and regression blocks. By default both of these sets of estimates are computed routinely.

A form comprising two dialogue boxes (one for each of the seasonal and regression estimates sets just described) is displayed. Enter **N** in a box to turn off the routine calculation of that set of estimates.

Estimate computation takes time, and storage requires computer memory. It is often the case in early exploratory analysis that individual factor estimates or regression block effects will not be required. This is the case, for example, when comparing forecasts for a selected model form using different priors, discount factors, or interventions.

• **Quit** The **Quit** command returns you to the Root menu.

15.21 Forward Intervention

The Forward Intervention menu provides subjective user intervention facilities at on-line analysis suspension points—monitor break points (if monitoring is turned on) and forward interventions (which are set using **Outlier-Intervention** command of the Model/Interrupts menu). If a monitor signal is generated at, say, January 1987, then intervention is made from the standpoint of having seen January's outcome, but not yet having reached February. A forward intervention point set for January 1987 causes analysis to be interrupted after December 1986 has been processed but before January 1987 is reached. The following sections describe the intervention facilities.

• **View Priors** The **View Priors** command allows you to examine the model based priors for the intervention time. See the **View Priors** command on the Forecast Priors menu.

• **Edit Priors** The **Edit Priors** command allows you to alter the model based priors for the intervention time. See the **Edit Priors** command on the Forecast Priors menu.

- **Interrupts** The **Interrupts** command is similar to the **Outlier-Intervention** command of the **Model/Interrupts** menu. The only difference is the range of times for which outliers and forward interventions may be (re)set. It is not permitted to change the status of times already included in the on-line analysis, so only those times not yet reached are included in the table presented. For example, if the data set is monthly, extending from April 1987 to October 1991, and an intervention is set for June 1989, then at that June intervention the **Interrupts** command will allow you to set outliers or forward interventions for July 1988 to October 1991 only.

- **Predict** The **Predict** command is similar to the **Predict** command on the **Root** menu. The difference here is that predictions are made from the standpoint of the intervention time rather than prior to any analysis or subsequent to the complete analysis of the data. The default values for the model component priors are the priors currently computed by the on-line analysis at the time of intervention. The priors may be changed—without affecting the values that will be used on resuming the on-line analysis—see the **Forecast Priors** menu above.

- **Abort** The **Abort** command terminates the on-line analysis, deleting all model estimates, and returns you to the **Root** menu.

- **Return** The **Return** command signals that you have finished all intervention activities and returns you to the on-line analysis. The pre-intervention forecast graph is restored and analysis resumes (with modified priors if changes were made during the intervention).

15.22 Monitor Signal

When any of the forecast monitors are operational and a breakdown in forecasting performance in the on-line analysis is signalled, a dialogue box appears describing the three available response options. Options are selected by pressing the indicated numeric key. The following sections describe the alternatives.

- **1 Ignore** The **Ignore** command instructs the program to ignore the current monitor signal, to reset the monitor to the starting state (the model is adequate, with a Bayes factor of 1), and to proceed with the on-line analysis as if no monitor signal had been generated.

- **2 Automatic** The **Automatic** command instructs the program to invoke the built-in rules for responding to a monitor signal.

  o **Rule 1** If the signal run length is 1, a single discrepant forecast-outcome pair, treat the offending response series value as an outlier and use the alternative set of component discount factors to compute priors for the next time point. For example, if February 1987 is signaled, then the prior for March 1988 is constructed using the exception discounts—which are typically much smaller than the routine discounts and therefore result in more uncertain priors so that existing (pre-February) information is given much less weight as the analysis proceeds.

  ○ **Rule 2** If the signal run length is greater than 1, use the exception discount factors to compute the priors for the next time point. Do not ignore the response value at the time of the signal. (Exception discounts are defined using the `Monitor` command on the Configuration menu.)

 • **3 User** The `User` command instructs the program to provide access to the subjective intervention facilities—see the Forward Intervention menu.

Index

Abraham, B., 177–178, 198, 230
Adaptive factor, 35
Advert, 123
Aggregate data, 35, 366
Aitchison, J., 75, 80
Alternative model, 21
Ameen, J., 92
Aoki, M, 89
AR, see ARIMA
ARIMA, 7, 83–89, 165–168, 178
ARMA, see ARIMA
Aswan dam, 169
Autocorrelation, 88, 125, 133, 136, 150, 155, 167
Autocovariance, 87
Autoregressive integrated moving average, see ARIMA

Back filter, 24, 270, 287
Back shift operator, 87
Balke, N.S., 165, 168–169, 230
BATS, 26, 96, 130, 144, 165, 235–238, 248, 250, 260, 280
Bayes factor, 62, 64, 66, 68, 312
    cumulative, 64, 66
    local, 66, 68
Bayes' theorem, 10, 16, 22–24, 30, 34, 40–41, 46, 56, 62
Bayesian
  analysis, 3, 9, 13, 16, 22, 236, 341
  forecasting, 10
  inference, 10
  learning, 16, 17, 22, 30
  methodology, 3, 11, 17
  updating, 34
Bazza, M., 230
Belief, 24
Berger, J.O., 27
Block
  discounting, see discount
  structuring, 26, 51
  updating, 70
Box, G.E.P., 69, 75, 80, 84, 88–89, 96, 230
Broemeling, L.D., 69

Candy example, 260
Carlstein, E., 165, 230
Carpenter, S.R., 230
Causality, 3
Change point, 168
Cholesky factorisation, 39
Cincinatti Bell, 180
Cobb, G.W., 165, 169, 230
Component,
  discounts, 129, 148
  estimate, 115
  intervention, 317
  model, 5, 9, 24–26, 51, 124, 261, 286
  prior, 105, 110–111, 115, 297
Conditional probability, 9
Configuration file, 258
Conjugate analysis, 55
Correlation, 274, 280
Cosine, 47, 50
Covariance, 19, 30, 34, 55
Cox, 96
Crude oil, 191
Cycle, 5, 24, 44, 47, 278, 286

Data transformation, 91-92, 119
  centre, 127
  difference, 87
  lagged variable, 142, 145
  moving average, 83–84, 186
  standardise, 128, 280
Death rates, 192
Decision theory, 20

405

Degree of belief, 17, 21
Degrees of freedom, 55–56, 262, 266
DeGroot, M.H., 27, 75, 80
Dialogue box, 248, 256, 262, 281, 293
Discount factor, 23, 47, 52–55, 59, 92,
    94, 96, 101–102, 119, 124, 129, 166,
    260, 265–266, 271, 278, 294
  block, 52–53, 124, 142
  component, 129, 148
  menu, 265
  optimal, 278
Discounting, 23, 59
Distribution, 17, 20, 24, 56
  conditional, 37, 56, 77
    forecast, 34, 56
  marginal, 60, 77
  multivariate
    normal, 56, 76
    normal-gamma, 80
    Student-t, 79
  skewed, 101
  unconditional, 58
  univariate
    gamma, 55, 59, 78
    normal, 15, 19–20, 24, 30–34, 36,
      37, 55–56, 75, 143
    normal-gamma, 79
    Student-t, 60, 78
DLM, 5, 8, 11, 13–15, 19, 22, 24–26,
    29–30, 37, 55, 58, 62, 92, 102, 236,
    259, 286
Dummy variable, 168
Dunsmore, I.R., 75, 80
Durbin, J., 301, 340
Dynamic linear model, see DLM
Dynamic model, 94, 96–98, 271, 278
Dynamic modelling, 16, 237
Dynamic regression, 143, 276
Dümbgen, L., 165, 230

Economic theory, 123
Eire, 92
Error analysis, 68
Estimation, 14
Evolution, 14–15, 23, 38, 47, 52, 59,
    94
  equation, 17–18
  variance, 17, 32, 70, 264
EWMA, 6, 7
Exception, 21

discount, 312–313
Expectation, 19
Expected value, 32, 58
Explanatory variables, 4
Exponentially weighted moving
    average, see EWMA
External information, 6, 17, 21, 60

Factor estimate, 290
File mask, 257
Filtering, see retrospective
Fitted estimate, see retrospective
Flu, 192
Forecast, 9, 18, 20–21, 31, 37, 262,
    263, 283, 295
  cumulative, 19, 32
    assessment, 20
  distribution, 18, 20–21, 31–32, 56,
    62
  error, 35
  function, 88
  horizon, 263, 283
  joint, 17, 36, 56, 62, 76
  likelihood, 62, 130, 133, 136
  mean, 32, 264
  menu, 265
  model, 9
  monitoring, 62, 130, 306
  performance, 20, 23, 62, 66, 68, 92,
    97, 102
  system, 20
  unconditional, 59
  variance, 23, 32, 295
Forecasting, 3–5, 10, 19, 26, 119, 140
Fourier representation theorem, 49,
    292
Freight volume, 186
Fundamental harmonic, 292

Gas consumption, 170
General Motors, 198
Greece, 147, 158
  Athens, 5, 158
  Aegean islands, 158
  Crete, 158
  Epirus, 158
  Ionian islands, 158
  Macedonia, 158
  Peloponnesos, 158
  Thessaly, 158

Thrace, 158
Griffiths, W.E., 230
Growth, 5, 92, 94, 105, 108
   quadratic, 44
   see also trend component

Harmonic analysis, 47, 50, 292
Harmonic model, 49, 51, 119, 286, 293, 294
Harrison, P.J., 15, 20, 27, 30, 35, 44, 47, 50–51, 58, 62, 65, 66, 70, 75, 80, 89, 92, 173–174, 191, 230
Harvey, A.C., 89, 301, 341
Hill, R.C., 230
Housing starts, 177

Independence, 19, 30, 34
Industrial sales, 338
Inflation, 174
Information, 8
   discounting, 22-23, 58
   filtering, 24
Innovation, 34
Input-output menu, 280
Intervention, 9–10, 17, 21, 26, 37–38, 54, 60, 104, 108, 111, 130, 148, 167, 301, 317, 325
   analysis, 131, 134, 144
   indicator, 313
Irregular time scale, 341

Jacobian, 96, 99
Japan, 163
Jeffreys, H., 64, 69
Jenkins, G.M., 69, 84, 88–89, 230
Johnson N.L., 75, 80
Johnson & Johnson, 194
Joint probability, 101
Judge, G.G.R., 230

Kalman filter, 84
Kotz, S., 75, 80

Lathrop, R.C., 230
Leap year, 147, 156
Ledolter, J., 177–178, 198, 230
Lee, T.C., 230
Level component, 105, 124, 259

discount, 278
   see also trend component
Likelihood, 16, 22–23, 30, 34, 96
Linear Bayes, 62
Log-likelihood, 96, 101
Lütkepohl, H.., 230

MA, see ARIMA
Mad, see mean absolute deviation
Management by exception, 10, 21, 95, 102
Market share, 121
Mean absolute deviation, 96, 130, 133, 136
Mean response function, 61
Mean squared error, 96, 130, 133, 136
Memory, 123, 145
Menu hierarchy, 245
Missing value, 3–4, 35, 236, 341
   BATS, 342
Model, 5, 9, 15–16
   building, 9
   choice, 9
   comparison, 64
   decomposition, 291
   likelihood, 34
   menu, 265
   misspecification, 129
   selection, 20
Modelling, 3, 8–9, 13, 16
Monitor, 64, 65
   sensitivity, 312, 331
   signal, 20-21, 65, 119, 131
   response, 21, 314
   run length, 310–311, 314
   system, 21, 26, 62, 103, 301
Mse, see mean squared error
Multiplicative seasonal, 191
Multiprocess model, 11

Nicholson, A.J., 231
Nile, 165, 168
Nonlinear model, 11, 24, 92, 119
Nonnormal model, 11

Observation
   equation, 15, 18, 25, 26, 30–32, 35, 45, 55, 56, 124
   variance, 15, 22, 35, 55, 260, 266

On-line
  analysis, 24, 37, 116
  estimate, 271, 277, 280, 290
  fit, 268
One period forecast, 19, 267, 276
OPEC, 191, 317
Ord, K., 89
Outlier, 111, 135
  indicator, 313

Pankratz, A., 89, 178, 180, 182, 184, 231
Passenger cars, 198
Period, 45, 47, 286
Periodogram, 295
Permanent income hypothesis, 184
Permutation matrix, 48
Phase, 47, 50
Pole, A., 44, 69, 89
Polynomial model, 42, 259
Post-intervention prior, 60
Posterior, 16–17, 23, 35, 37, 56, 59, 70
  distribution, 16–17, 22, 34, 37, 46, 56, 59
  information, 17, 30, 35, 59
  mean, 35, 283
  variance, 22
PostScript, 239–240, 256
Powerlaw, 58
Precision, 22
Prediction, *see* forecast
Prior, 16–18, 21–23, 32, 35, 37, 55, 56, 59–60, 95, 101, 104–105, 262, 263
  belief, 17, 35
    vague, 144
  distribution, 17, 46–47, 59
  information, 17–18, 30–34, 59, 95
  marginal, 56
  moment, 22, 35, 264, 284
  specification, 17, 131, 148, 297
Probability, 9, 10, 16, 20
  belief, 30
  density function, 30
  distribution, 9, 18–20, 22, 101
  statement, 24

Quantile plot, 136

Random walk, 15, 52, 70, 124
Reference analysis, 267, 276, 286, 294, 297
Regression, 3, 5–6, 8, 13, 15, 18, 26, 51, 61, 121, 124, 129, 275, 295
  block, 284
  coefficient, 124, 134, 136, 275, 277, 280
  component, 35, 51, 139, 277–278, 280
  discount, 278, 284
  model, 280, 283
  multiple, 130, 142, 284
  vector, 26, 30, 49, 51, 124, 260
Residual, 125
  analysis, 69, 150, 155
  fitted, 69
  forecast, 69
Response, 124, 261, 262, 275, 282, 286
Retail price index, 174
Retail sales, 172
Retrospective analysis, 24, 26, 60–61, 110–111, 115, 116, 119–120, 267, 269, 271
  distribution, 24, 40, 60–61
  estimate, 61, 111, 115–116, 267, 271, 277–279, 288
Rojstaczer, S., 231
Root menu, 244, 266–267

San Lorenzo Park, 6
Savings rate, 184
Scale factor, 55–56, 59
Scale parameter, 58, 61
  evolution, 62
  see also observation variance
Schizophrenia, 182
Seasonal
  amplitude, 47, 50, 92
  component, 6, 25–26, 35, 44–46, 286, 290, 292, 297, 301
    free-form, 286, 292, 294
    restricted, 292
  cycle, 6, 286, 293
  difference, 87
  discount, 290
  effect, 44, 291, 297
  factor, 44, 287–288, 291

pattern, 93–94, 105, 110, 286, 290, 294, 304
peak, 47
period, 293
prior, 46
trough, 47
variation, 259, 286, 290, 295
Sequential analysis, 16, 24, 26, 70
Serial correlation, 129, 143
Share earnings, 194
Shumway, R.H., 89, 192, 194, 231
Sine, 47, 50
Skew, 101
Smith, A.F.M., 58, 70
Smoothing, *see* retrospective
Standard & Poor, 198
State, 14, 30–31, 34, 37, 59
    estimate, 35
    forecast, 32
    parameter, 30, 56
    posterior variance, 35
    vector, 14, 26, 30, 36, 44
Static analysis, 280
Static model, 7–8, 14–15, 96, 98, 264, 265, 271, 280, 294
Stationarity, 86, 88, 295
    weak, 86
Steady model, 259, 262, 266, 278, 283
    see also trend component
Stevens, C.F., 58, 173, 230
Stream flow, 6-7
Structural change, 17, 66, 94, 103, 119–120, 149, 278, 329
Stuart, A., 89
Subjective intervention,
    see intervention
Subjective judgement, 17
Sulphur dioxide, 192
Superposition, 51

System equation, 14–15, 17–18, 22–23, 26, 30, 32, 35, 37–38, 55, 58–59, 124
    variance, 22, 260

Telephone calls, 180
Temporal adjustment, 8
Tiao G.C., 75, 80
Time scale, 3, 4
Time series, 3–4, 7, 14, 16, 24
    analysis, 3–4, 24, 237
Transformation,
    see data transformation
Trend component, 5, 24–26, 35, 42, 58, 94, 262, 264–265, 269, 275, 278, 286, 301
    constant, 70
    discount, 271, 290

U.K. gas consumption, 6
U.S. Geological Survey, 6
Uncertainty, 9, 10
Updating analysis, 37

Value added tax, 174
Vandaele, W., 89, 231
Variance component, 19–20, 31–32, 56, 59–60, 101, 266
    discount, 26, 61, 265
    law, 5, 58, 301
    learning, 22, 60, 260
Volatility, 21

Wei, W.W.S., 89, 231
West, M., 15, 20, 27, 30, 35, 44, 47, 50–51, 58, 62, 65–66, 69, 75, 80, 89, 191, 231
What if?, 320
What happened?, 270
Wolf, S., 231